Proof and Disproof in Formal Logic

OXFORD TEXTS IN LOGIC

Books in the Series

1. Shawn Hedman: *A First Course in Logic: An Introduction to Model Theory, Proof Theory, Computability, and Complexity*

2. Richard Bornat: *Proof and Disproof in Formal Logic*

Proof and Disproof in Formal Logic

An introduction for programmers

Richard Bornat
Middlesex University, UK

Great Clarendon Street, Oxford OX2 6DP

Oxford University Press is a department of the University of Oxford.
It furthers the University's objective of excellence in research, scholarship,
and education by publishing worldwide in

Oxford New York

Auckland Cape Town Dar es Salaam Hong Kong Karachi
Kuala Lumpur Madrid Melbourne Mexico City Nairobi
New Delhi Shanghai Taipei Toronto

With offices in

Argentina Austria Brazil Chile Czech Republic France Greece
Guatemala Hungary Italy Japan Poland Portugal Singapore
South Korea Switzerland Thailand Turkey Ukraine Vietnam

Published in the United States
by Oxford University Press Inc., New York

First published 2005

British Library Cataloguing in Publication Data

Data available

Library of Congress Cataloguing in Publication Data

Data available

Typeset by SPI Publisher Services, Pondicherry, India
Printed in Great Britain
on acid-free paper by
Biddles Ltd., King's Lynn

ISBN 0-19-8530269 (Hbk) 9780-19-853026-8
ISBN 0-19-8530277 (Pbk) 9780-19-853027-5

10 9 8 7 6 5 4 3 2 1

ReadMe.Preface

All sorts of people are interested in logic. Most are looking for ways to keep their thoughts straight, hoping for hints about mental hygiene and help with settling arguments. Only a minority are interested in **formal** logic, a 20th century development which allows you to check a logical claim without considering what the claim means. This highly abstract idea, once intended to be a sound foundation for mathematics but thrown out as inadequate, has found a home as an essential and practical part of computer science.

Mathematicians and philosophers study formal logic and worry at mathematical and philosophical problems: does it correspond to any interesting mathematics? does it characterize any interesting patterns of argument? Computer scientists, on the other hand, use logic as a tool: they **make** and **use** formal logical proofs. That alone would make it worth studying in the computer science curriculum. But there is more: the idea of a **formal system** — a collection of rules and axioms which define a universe of logical proofs — is what gives us programming languages and modern-day programming. All the hard and soft machinery that drove the information revolution through the second half of the 20th century and on into the 21st stands on the ideas of formal mathematical logic. Formal logic is **that** important.

This book concentrates on practical skills: making proofs and disproofs of particular logical claims. The logic it uses, called Natural Deduction, is very small and very simple. Working with it helps you see how large mathematical universes can be built on small foundations. It is a striking example of how it is possible to build useful things, even apparently meaningful things, on top of a meaningless formal system. It teaches you how to focus on **syntactic** reasoning, and turns you away from the **semantic** reasoning that dominates mathematics outside computer science.

If you don't know anything about logic...

...then skip the next few paragraphs. You can read them again after you've worked through the book, and they will explain what's happened to you. Start reading again where it says 'Jape'.

If you already know something about logic...

...then this book may surprise you. First, it starts with **proof** (most books start with truth tables). Second, it focusses on **constructive proof** (most books stay strictly classical). Third, it's **practical** (most books are foundational).

Those aren't quixotic decisions. If you want to **do** stuff with logic, making proofs is the most important skill, and proof is the obvious place to start. If you start with proof, constructive proof is easier than classical proof and it makes at least as much sense. Then the distinction between constructive and classical proof is subtle, important, understandable and quite, quite unbridgeable. Which is fun.

You always pay for your fun in the end, though. Classical models (truth tables) are easier to deal with than constructive models (Kripke trees). But constructive models are more interesting because they deal with the fun bit, the gap between constructive and classical logic where the controversial claims lie. So constructive models are fun too.

Last of all, because this book was devised for computer scientists, it shows how to use proof in anger. If you can program a bit, and if you wonder, as I used to, why your programs sometimes loop for ever, sometimes finish without achieving anything, or sometimes go along quite nicely until they fall splat! off the end of an array, then Hoare logic is for you. For most people it is a double eye-opener: first you see how hard it is to get programs right; then you see that with the aid of logic it is possible to get them right. Now **that's** real fun.

Jape

I began to get interested in formal reasoning when I was trying to teach people to program. I knew that expert programmers could justify and defend their designs, and I imagined that teaching novices to reason about programs would help them to learn how to program. After I was forced to recognize that I was wasting my time — no matter what your approach, people seem to teach themselves to program or not learn at all — I began to get interested in formal reasoning itself. Bernard Sufrin and I, two formal reasoning novices, got together and started to build a proof calculator to explain logic to ourselves. Bernard called the program Jape and the name stuck.

We always intended Jape to be a teaching **and** research tool, and it is, but it's had most success in teaching. I've used it to support this book. You can get Jape for yourself from the website www.jape.org.uk. There are versions for the major operating systems (Windows, MacOS X, Linux, Solaris), and you can find updates, news, and an address to write to if you want to suggest improvements or just complain.

Jape is a calculator. It deals with logic in the same sort of way that arithmetic calculators deal with numbers: you choose a button and press it; Jape makes the corresponding calculation step and shows you the result, but it doesn't give any advice about choosing steps or criticism about the step you chose. By trying out steps and then undoing if they don't work, you can use Jape to search for logical proof. Once you get good at search, you will find that you have learnt a **proof strategy** — a way of tackling problems that more or less guarantees success — and you can transfer that strategy to blackboard-and-chalk or pencil-and-paper or musing-on-the-bus proofs.

Even when you are skilled, Jape can be useful in checking a proof idea, because if Jape says it's a proof, then it certainly is. At every stage you can learn about the logic by reading the proofs that Jape helps you to make, and trying to see whether they justify the claims that they seem to prove (that's an example of **reflection**, which means no more than 'thinking about what you know').

Proof is the first and last third of the book. Disproof — the middle third — seems to be trickier than proof. But in practice, with Jape to help you, you can deal with it. Jape can help you calculate disproofs; you can use it to check if your latest idea is a disproof or not; you can use it to explore attempts and to explain why they are — or why they're not — valid disproofs.

Jape has one drawback: it's too much fun to use. It gives lots of positive feedback for not very much effort, and like any computer game it reveals to anybody who enjoys blasting away on the buttons just how easy it is to win. It is even possible (there's no secret: if I didn't tell you you'd find out anyway) to make proofs without really knowing what you are doing. Sometimes you can stumble on disproofs in the same way. That's fun to start with, but in the end it isn't enough, just as you can't live on only sweets, or only beer, or no sleep at all. So long as blasting the buttons is fun, Jape is probably doing you good. When you find the game is getting a bit of a drag, it's probably time to back off and learn some logic. You might decide to learn by using Jape, even!

Trajectory

This book divides into four parts, one small and three large.

- Part I: basics. Introduction to the idea of formal logic, via a short history and explanations of some technical words.
- Part II: formal syntactic proof. How to do calculations in a formal system where you are guided by shapes and never need to think about meaning. Your experiments are aided by Jape, which can operate as both inquisitor and oracle.
- Part III: formal semantic disproof. How to construct mathematical counter-examples to show that proof is impossible. Jape can check the counter-examples you build.

- Part IV: program specification and proof. How to apply your logical understanding to a real computer science problem, the accurate description and verification of programs. Jape helps, as far as arithmetic allows.

Acknowledgements

This book is the result of a long journey. Many people that I met along the way have helped me. I specially acknowledge those who were at various times my colleagues at Queen Mary College, including Abdul Abbas, Samson Abramsky, John Bell, Mark Christian, Keith Clarke, Mike Clark, Doug Goldson, Wilfrid Hodges, Peter O'Hearn and Steve Reeves. David Pym and Mike Samuels taught me everything I know about semantic modelling. Paul Taylor, Adam Eppendahl and Jules Bean improved my understanding and my presentation immensely. Adam de-gendered the mnemonic in Chapter 6 for me.

A special gold-embossed and heartfelt thankyou has to go to Bernard Sufrin. Bernard and I sat down together in 1992 to design and build Jape. Without Bernard's early design insights and his active collaboration over several years neither Jape nor this book would ever have existed.

Roy Dyckhoff's MacLogic was a major inspiration in the design of Jape, as was the design of the marvellous Macintosh GUI (and, of course, this book was written on a Mac!). Mark Dawson's thesis taught us about the sequent calculus, and the ideas pioneered by his supervisor, Krysia Broda, inspired us to try to build a calculator.

Thanks to Middlesex University for providing me with a haven when I cast myself adrift, and for supporting me while I finished this book.

This book was composed in Latex using TeXshop, making heavy use of the semantics package of Neergard and Glenstrup to sort out the equations, Jules Bean's boxribbonproofs package to compose the box-and-line proofs, and Tatsuta's proof package for the inference rules. I used `rtf2latex2e` to translate my original lecture notes.

The mistakes in this book are my own.

ReadMe.Teach

This preface is addressed to teachers.
Timid student readers may die of horror. Such persons should skip to Part I and start reading there.

As I said earlier: sometimes it seems that at university you can't teach anybody anything, and they just have to learn for themselves.

Logic is different. This book **works** (well I would say that, wouldn't I? — but it does!). Proof and disproof are **practical** skills, given Jape to help. An one-semester introduction-to-logic course can be a largely problem-driven experience for the student. I lectured Basics and Proof (with the exception of Chapter 4, which was covered in lab sessions) in three weeks or so, then gave three weeks over to exercise practice, followed by a test; then a similar but shorter treatment of Disproof, also with exercise classes and a test; then I showed a bit of Hoare logic to the keen ones and let the rest revise for a final test. I needed good lab assistants throughout, and luckily I always had them. (The material in Chapter 4, in particular, was the idea of Jules Bean and Mike Samuels. Thanks, guys!)

The results were gratifying (non-CS teachers look away now!): over 70% of an average first-year English university class learnt to do proofs reliably on paper and on blackboard; well over 50% could do disproofs in the same way. It's all helped by the fact that they can run Jape on their own computers, and Jape is quite fun to play with, so they do use it.

The Hoare logic part of this book is new. I've given definedness a central rôle, instead of sweeping it under the carpet as is sometimes the case. It's all very elementary, but (again, given the help of Jape) I believe it will be accessible.

Finally, I had limited aims. I didn't want to swamp novices with too much information, so I haven't tried to be encyclopaedic and I've tried not to go off on too many tangents. My aim is to tempt students into the logical forest; once they're in, surely we've got them! Someone who can do proof and has a notion of what a model is ought to be an easy touch for deeper logical ideas. Someone who can do program proofs might even start to reflect on what computer science is about. At the very least, somebody who has fun with logical proofs might stick around to listen. That's why I left out so much that you may think is essential to an introductory logic text. I hope this book will help you to have an eager and receptive audience when you add in all that stuff in later courses.

Contents

Part I

Basics

‘Formal logic’ is not a phrase that attracts. Everybody, perhaps, would like to be logical — but not **too** logical, in case they become unfeeling like Star Trek’s Mr Spock or antisocial like Viz’s Mr Logic. Hardly anybody wants to be formal: the word brings up images of stuffed shirts, stiff collars, exclusive people dressed up for expensive occasions that you and I can’t get into. We’d rather be informal, casual, easy, and logical only up to a point.

But formal logic is hot stuff, because it is the machinery in the engine room of computing. Computers do very simple formal logical reasoning, and can’t do anything else. The programming languages that drive those computers are formal logics. The protocols that drive the internet and the grid are formal logics too. Computers do what they do as well as they do because we know quite a lot about formal logic, and they keep falling over because we don’t yet know quite enough.

Chapter 1 gives some of the history of formal logic. It’s an easy read and it’s useful background. Chapter 2 introduces the language we use to talk about proof, disproof, reasoning and so on. If you skip it you will only have to come back to it later ...

1 A rough history of logic

You don't need to know history in order to understand modern formal logic, but a glimpse of history can help you to understand why formal logic is the way it is, and how it fits with the study of computer science. So I begin with a historical tale.

I've had to simplify the story considerably, and simplifications always distort. You can learn more from all kinds of sources: there are paperbacks on the main protagonists, there are online encyclopaedias, there are libraries. But you won't be misled if you believe it the way I tell it, even though I do simplify outrageously, and leave out volumes of interesting and relevant information.

1.1 The Greeks invent the game

The origins of modern formal logic are in Ancient Greek culture. In Athens, more than two thousand years ago, philosophers and mathematicians struggled with the problem of defining **valid arguments** — reasoning that would convince any rational person who attended to it. Judges wanted to be able to distinguish right from wrong and to do so reliably: they wanted to hear only valid arguments.

Very early on it was recognized that a persuasive **argument** falls into three parts. You start from accepted **premises** and use plausible steps of **reasoning** to reach a convincing **conclusion**. The ancients realized that if you accept the premises (I was in Sparta, and the crime was committed in Athens) and you agree that the steps are watertight (since nobody can be in two places at once, I wasn't in Athens; since nobody can commit robbery at a distance, I didn't do it) you are forced to accept the conclusion (I didn't do it!).[1]

Philosophers recognized that the problem is with the steps which connect premises to conclusion. They identified certain simple argument-shapes which never seem to mislead. They called these shapes **syllogisms**. The most famous example[2] is

[1] First outrageous simplification. Ancient Greek courts weren't like the ones we see in TV dramas, though they did have regard to truth.

[2] Another outrageous simplification: this is a modern syllogism from about a century ago. Ancient Greek philosophers wouldn't have recognized it, for various technical reasons.

All men are mortal;
Socrates is a man;
THEREFORE, Socrates is mortal.

which is an instance of a particular well-understood shape:

Every A is a B;
x is an A;
THEREFORE, x is a B.

This is a single-step argument. The premises — facts we have to agree before we can start — are on the first two lines; the reasoning is a single step to line 3, and reaches the conclusion immediately. (We'll see in Chapter 7 that in modern logic a similar argument takes more than one step.)

This argument-shape, like the other syllogisms the Greeks invented, appears to make watertight arguments no matter what we put for its **parameters** x, A and B. But it was noticed that even when we use a watertight argument, we don't always reach a convincing conclusion. A watertight argument shape will take us to a convincing conclusion if we start from accepted premises. But it can lead us far astray if we start from wild premises:

Every Martian is a cabbage;
Richard is a Martian;
THEREFORE, Richard is a cabbage.

Oh no I am not! I deny it! But I'm sure I'm not a Martian, so whether or not Martians are related to cabbages the argument doesn't persuade me that **I** am a cabbage. Contentious premises don't inspire confidence in the conclusion.

Weirdly, a watertight argument shape can reach an agreed conclusion from absurd premises:

Every cabbage is a man;
Richard is a cabbage;
THEREFORE, Richard is a man.

I do accept the conclusion, but I definitely don't agree with the premises. So am I a man or not? The premises are nonsense, but I don't have to reject the conclusion on those grounds. The argument shows that the conclusion follows from the premises, but since the premises don't correspond to reality, the argument is simply irrelevant, and it doesn't persuade me about anything at all.

What the philosophers decided was that a **valid** argument shape is one which will always take you from agreed premises to a convincing conclusion — one that must be accepted, can't be denied. If the argument shape isn't valid, or if the premises aren't agreed, then all bets are off.

All logical arguments are knock-down convincing if the place you start from is a good place to stand, and if the steps you take are good steps. Otherwise they are flaky and shaky. So to defeat an argument which reaches a conclusion you would rather not accept, you challenge its premises and/or its steps.

1.2 Goals galore, but we're not watching

Between ancient Greece and modern times lots of mathematics and philosophy was invented — if you prefer, discovered — and debated. I'm not going to discuss any of it, though there is a great deal that could be said. Amongst everything else, logic was extensively developed in the mediaeval period in Arabia and in Europe, the idea of the **algorithm** was invented, and so was the differential calculus. Lots of good stuff, but not precisely relevant to this history. It was necessary, though, and useful: we stand on deep foundations.

1.3 Frege changes the rules

Gottlob Frege, a philosopher, began the modern study of logic in the 1870s CE. He asked an apparently simple question: 'how do we know the truths of arithmetic?' For example, how do we know that when $x > 1$, $x^2 > x$?

We know what wetness is by experience — i.e. by experiment. We've stood in the rain, we've jumped in the bath, and if we need to know if a liquid is wetting, we can always stick a finger it. We know what green looks like. We have felt pain and known happiness. We can recognize the taste of a potato, and the feeling of sun on our face.

By contrast, Frege reasoned, we don't know arithmetic by experience. We know it **rationally**, we have been persuaded of its truth by argument. We can't experience all the numbers above 1, so we can't know by experience that for each of them, their square is greater than the number itself. But somebody might say: take the inequality $x > 1$ and multiply both sides by x — a safe procedure when $x > 0$, and therefore safe in our case since $x > 1$ — and you derive $x^2 > x$. That's a rational argument which might persuade you of a particular arithmetic truth.

But it's easy to make arguments which are so long and complicated that it's difficult to be sure they are valid. Tricksters can show an argument that starts with sensible premises and derives a nonsense conclusion like $1 = 0$, smuggling in an invalid step such as division by zero without making a fuss about it. If we are to really, truly **know** the truths of arithmetic, we must be persuaded of them by very solid arguments. Solid arguments like those Ancient Greek syllogisms, which won't lead us into error. Solid **logical** arguments.

The syllogisms which he inherited from the Greeks and the mediaevalists weren't enough for Frege. He invented a great deal of mathematics to underpin his reasoning, mathematics which we now recognize as the **predicate calculus**, a version of formal logic.

The word '**formal**' means 'by shape' and I've already smuggled the idea of shape into the discussion of valid arguments above. A valid argument can be recognized by its shape, its **form**. Frege intended to prove that arithmetic had good logical foundations by starting from **axioms** which are immediately acceptable forms (for example you might choose $A = A$ as an axiom) and proceeding with purely formal (shapewise) argument to derive the consequential truths of arithmetic.

In inventing his calculus, Frege was living Leibniz's dream. Gottfried Wilhelm Leibniz, a very great mathematician whose life and achievements I casually passed over in the previous section, was perhaps the first to suggest that mathematical reasoning might one day be reducible to formal calculation (building in turn on the work of another great mathematician, al-Khwarizmi, who had invented the notion of formal calculation which every child now learns in primary school). In the 1660s Leibniz imagined, but couldn't build, a machinery of argument which would save mathematics from plausible but faulty reasoning. He dreamed that the symbols themselves would drive the argument. Frege began to make those kind of arguments.

He began by using the recently invented set theory. He wanted to use the mathematics of sets to underpin the mathematics of arithmetic, and to use logic to underpin the mathematics of sets. That was because arithmetic is largely based on counting, and set theory might reasonably be supposed to explain counting. He went along for thirty years or so, making great progress, and many very important mathematicians jumped on the bandwagon.

1.4 Russell kicks Frege in the knee

In the early 1900s CE Bertrand Russell was one of the philosopher/logicians working on Frege's problem. Russell noticed something wrong with set theory. As used by Frege, it contained a paradox: there were remarks you could make which were self-contradictory. Some 'sets' were so defined that things had to be both in and out of those sets. That's paradoxical: when you define a set you do so by defining the things that are in it, so things that are both in and out (or, which is the same thing, neither in nor out) break the basic notions of set theory.

Russell defined his paradox in terms of sets that are not members of themselves. Hardy, his colleague, had a neat description of it, which I've de-gendered for modern sensibilities.

There is a village in which there is a cook.
The cook feeds everybody who does not feed themself,
and only those who do not feed themselves.
Who feeds the cook?

A village is a set of people. Within the village-set there is a cook-fed set, those people who are fed by the cook. Is the cook in the cook-fed set or not? The only way to be outside the cook-fed set is to feed yourself. But the cook can't feed the cook (...only those who do not feed themselves...). So the cook can't be outside the the cook-fed set and must be inside. But then the cook would be feeding the cook, so the cook must be outside the cook-fed set after all. And so on and on, round and round in circles for ever. The village, so simple to describe, **can't exist**.

Russell wrote to Frege describing the problem, severely denting his confidence (Frege didn't publish his work till after his retirement). Russell went on to find a way to alter set theory in order to eliminate his own paradox. What Frege had been working with began to be described as 'naïve' set theory, and Russell's correction was incorporated into mainstream mathematics. It pops up in computer science as part of the theory of types, and the use of types in programming languages (classes in Java, for example) owes a lot to it.

For various reasons Russell stopped working on logic. But others carried the project forward.

1.5 Gödel blows up the stadium

It was obvious to all, after Russell's intervention, that it wasn't going to be easy to show that arithmetic is really and truly founded on just a few obvious fact-shapes and some obviously valid argument shapes. Then in the 1930s Kurt Gödel showed that it isn't just hard, it's **actually impossible**.

What he did was to show that **if** you could make a logic that dealt properly with arithmetic, **then** by using arithmetic you would be able to code up logical claims which refer to the logic itself. Some of the claims could even refer to themselves. Dangerous paradoxical claims like 'this statement is unprovable'.[3] If that claim is true, then the logical language supports at least one claim that is unprovable. That means the logic is **incomplete**. Mathematicians don't like incomplete logics: if one claim is left out, what else is missing? But worse: suppose the logic is complete. Then you could prove that the claim is unprovable — that is, you could prove that it is true. But then it's provable, so it doesn't state the truth. Paradox! — we can prove a false claim, and the logic is **inconsistent**. Inconsistency is much, much worse than incompleteness; we must never believe

[3] Another outrageous simplification, and not quite a real example. See *Gödel, Escher, Bach* by Douglas Hofstader for a marvellous treatment of this whole matter.

something and its opposite at the same time, because if we do, then according to our logical principles, as we shall see, **anything** is believable.

Gödel's proof, that any logic which explains arithmetic must either be incomplete or inconsistent, seemed to have stopped the game. The project which Frege began could never be finished. Most mathematicians left the ground right away, and the crowds have never come back.

1.6 Turing and Church play on

Frege, and those who followed him, had built some wonderful mathematics. If it couldn't deal with arithmetic as Frege wished, was it any use? You bet!

What Frege, Russell and the others had produced was a way of defining a logic as a **formal system**: a collection of basic **axioms** (starting points) and **rules of inference** (ways of building upwards from the axioms). After Frege and Russell we talk about logics (there are lots of different ones, each with its own axioms and rules of inference) and about Logic (the study of logics).

Computer Science was built on these foundations. Turing used the notion of calculation to describe what a universal computing machine had to be like, and the marvellous things it could do if it was built. Church developed the λ calculus as a formal treatment of calculation, and since then every programming language — Java, C++, Fortran, Prolog, Miranda and all the others — has a basic collection of axioms (its instructions) and rules for building up programs from axioms and other programs (choices, loops, methods, blocks, classes, ...). Every programming language is a formal system, a particular special logic, and every program in that language is an argument in that logic.

What we have to allow, because of what Gödel proved, is that the logics used in computer science are necessarily incomplete. Early on, Turing proved that it is impossible to write a computer program which can read any computer program, look at the input you are going to give it, and decide whether that program will produce a result when given that input. There are many other such **undecidability** results that prove there are questions which you may ask but which formal calculation cannot answer.

The fact that there are programs you can imagine but can't ever write didn't delay computer science for long. Gödel left behind a very large field — it's infinite even though it can never be the whole of mathematics — and we've been playing in it ever since.

1.7 The beautiful game

History shows that we do in computer science is, in a deep and important sense, **logical**. Programs are logical things. If they weren't, we couldn't build unthinking (formal) machines which can execute programs using mechanical (formal) rules.

Because programs are built up, Frege-formal-system style, from simple pieces using simple rules, we can ask, and sometimes answer, logical questions about them. Does this particular program, given that sort of input, always give us the right kind of answer? This is like Frege's question about arithmetic: we can't discover the answer by experiment, because there are just too many examples of 'that sort of input' and 'the right kind of answer'. Usually there is an infinite number of cases to consider.

We have to fall back on logical argument. But how do we say logical things about programs? How do we describe the input? How do we describe the output which we hope to see? How do we reason that the program we have before us actually does the right thing, once we've managed to say what it is we want?

The answer is: with some difficulty, but by and large using a logic derived from Frege's predicate calculus. Even if you never get round to proving anything about a program, predicate calculus and set theory can be used to describe what a program **ought** to do. Describing what ought to happen is called **specification**, part of software engineering, which is very big business indeed. To specify a program you have to have some experience of logical argument. Even if you specify in a language which isn't predicate calculus — if you use UML, say — you will be appealing to logical argument.

There are more ambitious uses of logic. We can program (with some limitations) **in logic itself** using a language like Prolog. We can build our programs by starting with a formal specification and using logical steps to **refine** an accurate program, as in Abrial's B method. We can look for ways to describe the world and the actions of a robot in logical terms, as in artificial intelligence research.

Even if you aren't a computer science theoretician; even if you never study any logic ever again; logic won't desert you. Did you ever hear a politician argue, and feel 'there's something wrong with that argument'? If so, you need logic as a nonsense-detector. We'll get to that quite soon, after I have laid some foundations.

2 How to speak and read logic

Education is about ideas rather than facts. The ideas of logic, and especially the ideas of formal logic, will be new and strange to most of my readers. New ideas need new words, new technical language. Outsiders sometimes call technical language 'jargon' and imagine that experts are hiding something by not sticking to everyday English. They are wrong: just as you can't discuss the design of folding bicycles with the same words that you use to discuss human biology, so you can't discuss formal logic without using an appropriate language.

Some of the words we use to discuss logic are common words, but we give them alternative meanings, or at least alternatively precise meanings. Those precise meanings really matter. When you talk about logic, you won't be able to say what you mean unless you use the right words — technical words, not everyday English words — and unless you use them carefully. You have to learn to **speak our language**.

No matter how hard the lexicographers try, nobody can quite write down the meaning of a word (try looking up 'right' and 'left' in a dictionary). Words have to be understood, their meaning acquired by use and by hearing others use them. To understand words we must be active — by speaking and writing — as often as, or more often than, passively listening and reading. This Great Mystery of Education is another reason to learn to speak our language.

Despite the difficulty of description and definition, I shall try to describe here some of the more important technical words that are used — and explained by that use — in later chapters.

2.1 Formal = by shape

The word **formal** means 'by shape', **by form**, rather than by content or by meaning. Outside logic it is often used as a term of social or political abuse. You can formally comply with the law by obeying its letter but not its spirit. Dancers at a formal ball must wear the correct clothes so that their outsides look the part, however villainous they might be inside. Informality, on the other hand, obeys no rules and can appeal to a deeper understanding than mere form.

Computing machinery is formal, because it can't be anything else. People are naturally informal, but can learn to be quite formal if they try hard.

2.2 Argument = line of reasoning

To argue, says the dictionary, is to seek to show by reasoning: hence an **argument** is a line of reasoning. Lawyers use the word in just this way. Outside logic and the law an 'argument' is a disagreement between people, a heated dispute. Our arguments, especially our **logical arguments**, should always be calm and measured — though if you eavesdropped outside a computer-science researcher's office, you wouldn't always think so.

2.2.1 Parts of an argument. A **premise** is the starting point of an argument. It's often expressed as a **hypothesis**, a supposition. 'Suppose that the universe contained only electrons' is a premise at the beginning of an argument, the position from which the conversation can continue.

A **conclusion** is the finishing point of an argument, what the argument 'proves'. 'Then we wouldn't be here to notice' might be the conclusion of the electronic-universe argument, for example.

Notice that the conclusion holds only if (or where or when) the premise holds. Logical arguments are often about **hypothetical** — supposed, often imaginary — situations. These are the kinds of argument which politicians usually (and sometimes less than honestly) refuse to entertain.

2.2.2 What to do in an argument. Three words — **refute**, **infer** and **imply** — are used carefully by logicians but often used inaccurately in the non-logical world. You **refute** an argument by showing that it is logically mistaken. People in trouble, though, sometimes say that they 'refute' an accusation when in fact they are **denying** it, simply stating that they don't believe it and that neither should you. Denial is easier than refutation, but refutation is stronger: refuting an argument demolishes it; denying is just shouting 'No!'.

You **infer** a conclusion by reasoning. Inference is something you can choose to do: you don't have to infer from my shifty appearance, the bloodstains on my coat and my presence at the crime scene that I am the murderer, but you may decide to do so. Your inference may be correct or incorrect: not every shifty-looking person is a murderer, not even the bloodstained ones. Maybe I'm just a particularly dishevelled detective.

You **imply** a conclusion if you provide evidence for it, but don't actually do the work to infer it. Implicit conclusions are the spice of political life. In the UK political scandals of the mid 1960s journalists had lots of fun with remarks like 'there is no truth in the rumour that the minister is a drunk', hinting that he might be but evading the libel laws by appearing to assert the opposite. Implications were sometimes quite subtly expressed, for example when a newspaper put pictures of prominent figures on the same page as a scandalous headline, but in a different story.

Inference can be jumping to conclusions; implication can be innuendo. Fights can start if somebody thinks your remarks imply an insulting conclusion, and chooses to infer that you meant it even though you didn't say it.

2.3 Proof and disproof

A **proof** is a test which, if passed, is a guarantee of some quality. It's an old word: people 'prove themselves' in sport or in battle; whisky was 'proof' if it burnt when it was thrown on the fire; the proof of the pudding is in the eating. We extract a technical meaning: a **mathematical** or **logical proof** is an argument which is so well made that it would persuade any intelligent reader. A **formal proof** is a logical argument which convinces by obeying formal (shape-wise) rules, and which can therefore be read and understood without referring to its meaning.

2.3.1 Scientific proof and disproof: by demonstration. In science and in everyday life a **proof** is a practical demonstration of the truth of a claim. You point to (demonstrate) something and say 'there's the proof'. You might claim, for example, to be able to run 100 metres in less than 10 seconds: you prove it to me by actually doing it while I time you. Showing me a piece of paper signed by an official timekeeper, or even an Olympic medal, isn't a direct demonstration, but if I decide to believe you then I can treat it as proof.

Scientific and everyday **disproof** is also a demonstration, showing that a claim is false. If you claim that all your apples are good, and I can show you a bad apple in your box, I've disproved your claim. If I claim that nobody can run 100 metres in less than 9 seconds, you can disprove my claim by running very fast while I watch.

Demonstration is really a scientific notion. By scientific standards, for example, it's well established that there are no wild wolves in England in 2004. Nobody's seen any droppings or paw prints or heard any howls for a hundred years or more. For scientists, and for most of the rest of us, that's evidence enough to amount to proof. It isn't logical, nailed-down dead certain proof, though. Nor would a disproof be: if you took me to a lonely spot on the moors and showed me a family of wild wolves, you could be fooling me with painted dogs or wolves you'd got from a zoo. I'd take a while to be convinced that the wolves have really come back. (Wild pigs, though: we've got them. They escaped from farms and trot about the woods at night. Really they do!)

Everyday proof and disproof is a demonstration of the truth or falsity of a claim by showing something real. But we all know how easily we can be deceived about what's real — by stage magicians, for example, by faulty measuring instruments or by fleeting glimpses. Proof by demonstration — scientific proof — is a slippery thing. Scientists work hard to find out what is and isn't true about the physical world, and are always sure that they have never quite found it. I'm equally sure that they are nearer to truth than anybody else could be.

2.4 Mathematical proof and disproof: by argument

Mathematical truths, if they exist, aren't a matter of experience. Our only access to them is through reasoned argument. Even an obvious truth like $3 > 2$ has to be demonstrated by appealing to notions of counting and one-to-one correspondence, requiring the audience to abstract from reality and to generalize from experience. Slightly more subtle truths, like 'when $x < 0$ or $x > 1$, then $x^2 > x$' require more subtle argument still.

For mathematicians, therefore, a proof is always a **convincing reasoned argument**. A really good proof is one which would convince anybody of its conclusion. Sometimes the argument can be really simple: if all that is required is a demonstration that we can make some sort of mathematical object, then we just describe how to make one. More often we have to make intricate arguments with many steps of reasoning.

2.4.1 Proof by the rules. I've glibly written, several times so far, about 'mathematical truths'. Mathematical truth is no less slippery a notion than scientific truth. It's not clear, as you may be surprised to find out in Chapter 3, that we ought to rest our ideas of logic on the idea of truth.

Ever since the ancient Greeks laid the foundations of the subject, logic has tried to define convincing arguments as those conducted according to rules. Formal proof is argument carried out according to formal rules. It's possible in this way to define precisely what we will accept as a convincing argument, at the cost of leaving out some arguments which we'd rather include, and of embracing some others which we'd rather exclude. In this book, a proof is **an argument which follows the rules** to establish a mathematical claim. You will see what 'the rules' are in Part II.

What we really care about in the real world is the usefulness of a proof. A proof-by-the-rules ought not to lead us astray. So, as well as being easy to recognize and easy to use, the rules ought to correspond to reality. By the end of this book you will be able to judge how well mathematicians have succeeded in that aim.

2.4.2 Disproof by counter-example. Since a proof is an argument which follows the rules, a mathematical disproof has to be an argument that no proof can possibly exist — an argument about possible arguments. That sounds as if it might be rather difficult, and indeed it usually is. But sometimes you can do it more easily, if all that is needed is a demonstration. A demonstration disproves a claim by showing an instance which doesn't fit, as we might disprove the claim 'all professors are cabbages' by pointing to a professor who evidently isn't a cabbage.

Instances which don't fit a general claim are called **counter-examples**. Just one counter-example destroys a claim; a hundred examples, or a thousand, or a million, can't outweigh it.

If we are going to disprove by counter-example, we need to be able to point to instances of things which don't correspond to a logical claim. That means that even if we rely on the notion of proof-by-the-rules our arguments have somehow to relate back to reality. Part III presents examples of what mathematicians call a **model** of reality that give meaning to proof-by-the-rules. The discussion in Chapters 3 and 6 appeals to notions of meaning to justify the rules, so proof and disproof are bound up together from the beginning.

2.5 Proof, truth and knowledge

The rules of a formal logic, for example the ones set out in chapters 3 and 6, are designed to make it easy to decide whether an argument really is by the rules or not. You can know when you have a mathematical proof which have followed the rules, and the existence of the proof then becomes a fact that you can demonstrate. Do you then **know** what the proof claims? Philosophers tell us that knowledge is justified belief: if I have a proof then surely it justifies me in believing its claim.

It's in this particular sense that the word 'know' is used in this book: I know X if I have a proof-by-the-rules of the claim X. I sometimes use 'believe' to mean that I think a proof is possible but I haven't made one yet.

If I have a proof of a claim I'm forced to accept the claim, whether I like it or not. If I want to reject a claim, I'd better have a disproof.

2.5.1 Claim = remark about the world. The remark 'go home!' is a command, not a claim. You obey commands; you can't prove or disprove them.

On the other hand, nobody can obey the remark 'it is Thursday'. It's a claim about the relationship between the current instant and the calendar. You accept or reject claims, not obey them. The remark 'you are a cabbage' is untrue and often offensive whenever addressed to a human, but is still a claim.

Claims are **declarative** remarks — they describe something about the state of the world or somebody's understanding of it. Commands are **imperative**: they are delivered in the expectation or at least the hope that they will be obeyed.

2.6 Basic logical principles

In the real world people try to use logic to get other people to accept a conclusion, hoping that they will then act on that acceptance. But logic is a slippery tool if what is needed is truth. Lawyers know this very well: they do their best, but all a court can do is reach is an understanding of who has the best argument. Logic was invented to help pick out good arguments from bad. It doesn't always work perfectly, but it's the best we can do.

Computers are mathematical machines. They have no choice but to calculate by rules — that is **all** they can do. The programming languages in which we instruct them are based on formal logical principles. So, in the restricted universe of computer science, logic can be made to work for us perfectly provided we stick to certain mathematical principles. Some of the principles are straightforward; others take a bit of getting used to.

2.6.1 Consistency. If you believe something and at the same time you also believe its opposite, you are surely confused. Mathematicians would say that your beliefs are inconsistent, and inconsistency is not a good idea. If you accept, for example, that today is Thursday and at the same time that it is Tuesday, your head is in a mess. If you really believe it's Thursday then you believe that it's the third day since Monday; if you believe it's Tuesday then you believe it's the first day since Monday. So you believe that $n = 3$ and at the same time that $n = 1$; that means, according to the normal rules of algebra, that you believe $1 = 3$ and we can lead you a merry arithmetical dance. You had better sort your head out before you decide what to do today.
The **consistency** principle is

Definition 2.1 *contr adictions arent allowe d.*

A contradiction is the simultaneous acceptance and rejection of some remark: if that describes your state of mind then, clearly, you are confused, your thoughts are inconsistent, and — according to the principles of logic — you can't reason properly any more.

2.6.2 Hypothesis. Logical arguments proceed from premises to conclusion by logical steps: in this book, by steps which follow formal rules. One way of understanding the rules is to say that they are about consistent states of mind. Each rule says: 'if you accept *that* claim (or those claims), then to be consistent, you can/must/should accept *this* claim'. This leads us to the notion of **hypothetical** proofs: arguments not about what is, but '**what if?**'. We aren't asked to accept that the conclusions of a hypothetical proof are actually true, only that they are a logical consequence of the premises.

Just as in the ancient Greece of Chapter 1, if you don't like the conclusion of a logical argument you must attack the premises and/or the steps. But you can't attack the premises: we aren't asking you to accept them, just asking you to consider what it would be like if you did accept them. It follows that our proofs have to be attacked on the steps of their arguments — that is, on whether they follow the rules or not.

2.6.3 Monotonicity. In the world of proof-by-rule, proofs can't be contradicted. A proof is a proof because it obeys the rules; the rules never change, so once you've made a proof it's forever. That's very unlike our real-life experience of argument, when new evidence can turn up which makes us change our mind, or a second, more careful, reading of an argument can overturn it.

Real-life arguments depend on evidence. If I can show that the premises of a real-life argument contradict reality, then I can dismiss the argument. (I might not find it so easy to reject the conclusion: a bad argument doesn't disprove a true conclusion!) In logic, on the other hand, the argument is the thing: we suppose the premises and examine only the reasoning.

In real life our arguments are rarely purely logical, particularly because we must often work from insufficient evidence. We'll only be sure about global warming if it actually happens; we may decide that we have to do something about it even though we can't be sure it will ever happen. Real life asks us to judge from uncertainty, logic deals with certainty.
The **monotonicity** principle is

Definition 2.2 *Adding more premises cant invalidate a proof.*

You might, for example, persuade me that if I accept A I must logically accept B. But what if I also accept C, which contradicts B? The monotonicity principle says that B is still a logical consequence of A, so you must still accept it, no matter what C says about A or B, even if it directly contradicts them. Once a proof, always a proof!

2.6.4 So what? Logic is a mathematical abstraction of real-world reasoning. Abstraction blurs distinctions, ignores some features in favour of others. One major simplification in logic is in the treatment of evidence. In real life we have to decide what to do based on what we know now; we always have at the back of our mind the thought that tomorrow we might find out something which contradicts the conclusion we've reached. We reason scientifically rather than mathematically, putting evidence above everything else, always ready (if we're honest) to change our mind.

Mathematical logic doesn't work like that: you imagine the evidence, the reasoning is what matters, and the conclusions once reached can't be changed. That makes it unlike everyday reasoning, but that doesn't mean that everyday reasoning is right and mathematical logic is wrong, nor vice-versa. It's just that abstraction ignores detail. In the case of logic, the messy details of living with partial and unreliable evidence have been abstracted away.

Formal logic arose out of the study of language and human reasoning. Out of formal logic arose computer science. For computer scientists, the connection between logic and everyday argument is not the point. Monotonicity and consistency are mathematical principles which make logic mathematically tractable. They aren't arbitrary principles: they are very old, very well-established, and most mathematicians would say that you **must** stick to these principles if you want to define **any** kind of mathematical logic. Computer science and computer programming and computer hardware certainly depend on them. We'll do the same.

2.7 Logical rules

A logical or **inference** rule is an abstraction of a step in an argument. You might tease a friend or abuse an enemy as follows:

> Your bottom is made of rhubarb, **and**
> your head is made of custard, **and**
> you are sitting in a bowl;
> **therefore** you are a pudding.

This argument is not likely to be frequently useful (unless you have a very peculiar set of friends) but you can still make an argument pattern out of it.

A logical **rule of inference** shows how to make a larger proof of some claim out of one or more smaller proofs of related claims. Here, for example, is a rule abstracted from the argument above:

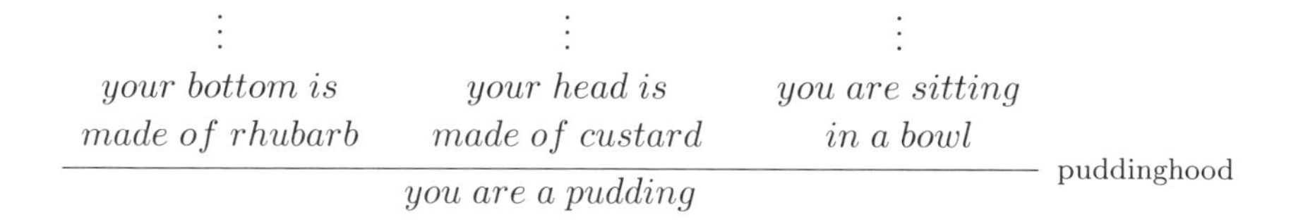

The claims above the line are the **antecedents** of the rule, and the claim below the line is its **consequent** ('antecedent' means 'going before'; 'consequent' means 'coming after'). The name of the rule is 'puddinghood', and it's written to the right of the line. The columns of dots above the antecedents show where you need to plug in **sub-proofs** that establish the antecedent claims.

Proofs don't usually take place in a vacuum: they start with premises. A premise really means 'suppose we had a proof of ... ' and we can call on it as if it was a proof. Implicitly each of the antecedent sub-proofs relies on the same premises. A sub-proof can be just a premise: if you assume my head is made of custard that's the end of the matter so far as the puddinghood step is concerned.

2.7.1 Forward and backward reasoning. One way to use the puddinghood rule is to use proofs or premises that you already have about rhubarb, custard and bowls to infer what you may think is an amusing pudding-insult. That's the **downward** or **forward** reading.

The other way to use the rule is to plan how to persuade me that I'm really a pudding. According to the rule, you can do that if you can find separate proofs or premises, one about rhubarb, another about custard, a third about a bowl. That's the **upward** or **backward** reading of the rule.

Forward readings say: if you already accept the stuff above the line then logically you must accept the stuff below the line. It's the way we generally read proofs once they are finished. Backward readings say: if you want somebody — even yourself — to accept the stuff below the line then logically you need only persuade them to accept the stuff above the line. It's often a good way to search for a proof of a claim.

The puddinghood rule tells us one way to prove that somebody is a pudding; that **doesn't** mean that it's the only way to do it. There are, for one thing, various different kinds of puddings. There is the possibility of indirect proof. Rules show one way that a proof **might** go, not the way that it **must** go.

2.7.2 Assumptions. I'd be offended to be thought a pudding, but you might try to argue that I shouldn't mind. I find the nice-pudding argument queasily unconvincing:

> 'Suppose you really were a pudding. Everyone would like you! And that would be nice. So being a pudding is nice, really.'

My problem is the slide from 'like' to 'nice': it isn't always nice to be liked, especially if the consequence is being eaten. But things may be different for puddings, after all, and in any case it's a good example of an argument which involves an **assumption**, an invitation to consider an alternative state of the world.

The nice-pudding argument is an instance of a general shape: 'Suppose A; you would then be pleased; therefore A is pleasing'. It seems to be a valid argument shape,[1] and it can be captured in a rule of inference.

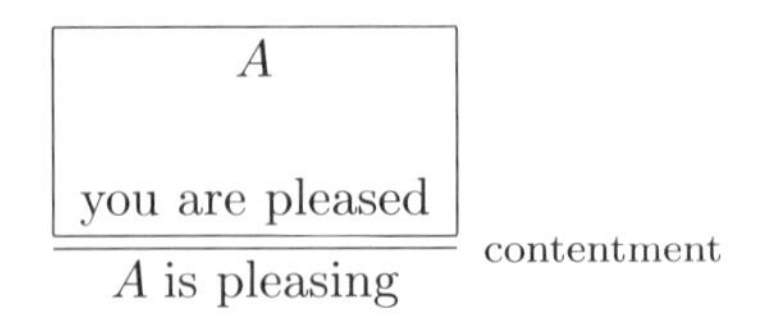

The three dots in the rule show where you have to put the argument which connects the **assumption** A to the sub-proof's conclusion 'you are pleased'. The

[1] It certainly is: it's a special case of the $\rightarrow$ intro step of Chapter 3.

box restricts the way that the assumption A can be used: it's valid inside the box, but not outside. The box isn't an absolutely tight enclosure, because claims can leak in: you can use a proof's overall premises inside the box whenever you need them. The boxed sub-proof introduces an additional hypothesis, which can be used inside the box but not outside.

2.8 What does logic mean?

This book is mostly about formal reasoning: reasoning by form, by formula-shape, without regard to what the formulae mean. That's a very strange thing to want to do, except as a mathematical game. It can be worth doing in practice only when it has a connection to reality and when it turns out to be easier than reasoning by by appeal to meaning. In the case of formal logic we have both those conditions. Evidence of the connection to reality is the existence of computing machinery which makes formal calculations which we find useful. Evidence of the ease of formal calculation is the speed of computers in coming to their formal conclusions.

Although they are used as if they are meaningless, in practice formal rules are invented and justified by appealing to the meaning of formulae. That isn't quite so comforting as it sounds, because the meaning we give to formulae is carefully thought out to fit the formal rules. This sounds like a rope trick, floating in the air without support: formal rules depend on meaning which is crafted to fit formal rules which depend on meaning ...

To break into the mathematical magic circle we must start with our feet on the ground. When describing rules in Chapters 3 and 6 I give an informal description of meaning for each formula shape; the description is carefully chosen to make immediate sense and at the same time to correspond as closely as possible to the more mathematical definition of Part III. Then I use the informal description to justify some formal rules of inference. Practice helps you understand how the formal rules work; then practice with the mathematical model closes the circle and you're flying.

Every logical formula makes a claim. Since this book is largely about proof-as-argument I give the meaning of a formula by describing the sort of situation in which an argumentative opponent can force you to accept the formula's claim. That isn't the only way to deal with meaning. I will discuss in Chapter 3 the relationship between meaning-by-argument and meaning-as-truth and what effect that has on formal reasoning.

2.9 Pronunciation

Something very odd happens when you say something you don't quite understand. Your mind revolts, as if you didn't really believe what you were saying, as if you were trying to tell a lie. Talking out loud about things you think you understand often reveals to **you** that you don't understand them. That's why conversation — tutorials, exercise classes, discussions with your friends — is so important in education. Sometimes the quickest way of finding out whether you understand an idea, as every teacher knows, is to try to describe it to somebody else.

You can use conversation to help you to understand logic, but you must remember 'I've gotta use words when I talk to you'. Strange logical operators that were only shapes on a page become real, become more understandable when you only speak their names. So: speak their names; talk to your friends about them. It's the only way!

2.10 Formal notation

This book makes use of some symbols which will be novel to most readers: the logical connectives $\wedge$ (and), $\rightarrow$ (implies), $\vee$ (or) and $\neg$ (not); the logical quantifiers $\forall$ (for all) and $\exists$ (exists); the symbols $\top$ (truth) and $\bot$ (contradiction); the turnstile symbols $\vdash$ (proves), $\models$ (models) and $\Vdash$ (forces). The meaning of the connectives and symbols is described in Chapter 3 and their use in arguments is explored in Chapter 5. The meaning of the quantifiers is described in Chapter 6 and proofs are explored in Chapter 7. Then Part III goes over everything again, this time with an eye to disproof. Finally, the notation is employed for real in Part IV.

I use symbolic names to stand for particular kinds of formula: A, B and C are 'formula parameters' in the inference rules of Part II and the definitions of Part III; E, F and G are simple formulas in examples; i is the name of an 'individual' in inference rules and definitions; j and k are names of individuals in examples; P is a 'predicate parameter' in inference rules and definitions; R, S and T are simple predicates in examples; x, y and z are 'quantified variables' in inference rules, definitions and examples.

To put it another way: in inference rules and definitions I use A, B, C, and $P(i)$; in claims I use E, F, G, j, k, $R(\ldots)$, $S(\ldots)$ and $T(\ldots)$.

In Part IV I change my conventions slightly: a, b, c are array program variables, and i, j, k are integer program variables.

Part II

Formal proof

This part of the book is about formal proof in a particular logical system, Gerhard Gentzen's simple and beautiful **Natural Deduction**, developed in 1935.

As a result of Gentzen's brilliant work, there isn't much to *know* about Natural Deduction — just a few symbols, each with two or three rules. I divide the presentation into two chapters: the **connectives**, which are like the operators of arithmetic, are described in Chapter 3; the **quantifiers**, which are like procedures, functions or methods in programming languages, are described in Chapter 6. Description is less than half the job: Chapters 4 and 5 show how to make proofs which involve the connectives, and Chapter 7 does the same for the quantifiers.

Reading about proof is not the point either: Chapters 5 and 7 show you how to make your own proofs of example problems.

3 Connectives

The connectives of logic are used to build larger claims out of smaller claims, just as the operators of arithmetic — $+$, $-$, $\times$, $\div$ and so on — make larger calculations out of smaller. There are only four connectives used in this book, shown in Table 3.1.

Table 3.1 Connectives of natural deduction

Connective	Simple name	Latinate name
$\wedge$	And	Conjunction
$\rightarrow$	Arrow, if-then	Implication, conditional
$\vee$	Or	Disjunction
$\neg$	Not	Negation

For each connective in turn I describe what a formula using the connective means, and then present inference rules which make those meanings work in formal proofs. The descriptions of meaning are informal, and based on the definition of circumstances in which a fair-minded person would feel forced to accept the claim made by a formula. Part III makes these descriptions more mathematically precise.

This chapter also introduces two symbols, shown in Table 3.2. Contradiction ($\bot$) is powerful and must be used sparingly; its meaning and use is bound up with the meaning and use of negation. Truth ($\top$) is universal but more or less useless, although it will get a little run out in Part IV.

Table 3.2 Truth and contradiction symbols

Symbol	Simple name	Latinate name
$\top$	Top	Truth
$\bot$	Bottom	Contradiction

3.1 Conjunction/and

Conjunction is used in situations where you accept more than one claim. The formula $A \wedge B$ is pronounced "A and B".

Definition 3.1 *When you accept $A \wedge B$, you are forced to accept both A and B.*

That is: accepting A and B together means you must accept them separately. Vice versa, when you accept them separately it is just as if you accepted them together.

You can use multiple conjunctions to build up larger collections of claims. $(A \wedge B) \wedge C$, for example, says that you are forced to accept $A \wedge B$ and you are forced to accept C; because of the definition of $A \wedge B$, it's clear that you are forced to accept A and you are forced to accept B and you are forced to accept C.

Although it doesn't matter, so far as meaning is concerned, how you bracket a conjunction — you have to accept each of the separate parts individually in any case — formal reasoning doesn't allow any ambiguity. If you write a multiple conjunction without using brackets then it's read as if you'd bracketed it to the left: $A \wedge B \wedge C \wedge D$, for example, is read as if you'd written $((A \wedge B) \wedge C) \wedge D$.

Insistence on the way you position brackets may seem a little silly at first, but in formal reasoning precise is precise is precise, as you will find when you begin to play with Jape. You even have to prove that $(A \wedge B) \wedge C$ means the same thing as $A \wedge (B \wedge C)$. Chapter 4 goes into more detail about why this kind of detail matters and even suggests how to pronounce brackets. Chapter 5 shows you how to make the proofs.

3.1.1 Reasoning with conjunction. Suppose I accept $A \wedge B$. You can point out to me that I'm forced to accept A. Or you can point out that I'm forced to accept B. That's how you **use** my acceptance: $A \wedge B$, therefore A; $A \wedge B$, therefore B.

Suppose that I accept A for some reason and simultaneously, for the same or some other reason, I accept B. Then, since I'm a fair-minded person, I surely have to recognize that it is just as if I accept $A \wedge B$. Logically, then, I can't deny $A \wedge B$. That's how you **persuade** me: A, also B, therefore $A \wedge B$.

This is all so simple it seems that there must be a trick, but there isn't. Conjunction is really straightforward. But we still have to be careful. $A \wedge B$ claims that you are forced to accept A and B separately, if called upon. It doesn't claim any more than that. In particular, it doesn't claim that there is any association between A and B, other than the fact that for the time being you accept them together and separately.

3.1.2 Rules for conjunction. Two **elimination** rules in Table 3.3 — so called because reading top to bottom they eliminate the $\wedge$ connective from a formula — capture the use argument. One **introduction** rule — so called because reading top to bottom it introduces an $\wedge$ connective into the proof — captures the persuasion argument. And that's it for conjunction!

Table 3.3 Rules for conjunction

$\dfrac{\begin{matrix}\vdots \\ A \wedge B\end{matrix}}{A}\ \wedge$ elim	$\dfrac{\begin{matrix}\vdots \\ A \wedge B\end{matrix}}{B}\ \wedge$ elim	$\dfrac{\begin{matrix}\vdots & \vdots \\ A & B\end{matrix}}{A \wedge B}\ \wedge$ intro

3.2 Implication/conditional/arrow

Implication is used in situations where you are asked to accept an association between claims. The formula $A \to B$ is pronounced "A implies B" or "**if** A **then** B" or, most simply, "A arrow B". The claim it makes is

Definition 3.2 *If you accept $A \to B$, then whenever you accept A you are forced to accept B.*

This definition captures the notion of implication as innuendo: acceptance of A drags acceptance of B behind it. The A claim, together with the $A \to B$ claim, conveys a claim of B without actually saying so. Logical implication also captures some of the meaning of 'leads to' and 'causes' — but imperfectly, as we shall see in Chapter 5.

Brackets matter in implications, unlike conjunctions: $(A \to B) \to C$ does **not** mean the same as $A \to (B \to C)$. Unbracketed implications bracket to the right — again, unlike conjunctions — so $A \to B \to C$ is read as if you'd written $A \to (B \to C)$: whenever you accept A, you are forced to accept $B \to C$. From that you can deduce that whenever you accept both A and B, you are forced to accept C. This suggests that $(A \wedge B) \to C$ ought to be a logical consequence of $A \to B \to C$ and indeed it is: you'll see how to prove it in Chapter 5.

3.2.1 Reasoning with implication. Suppose that I accept $A \to B$. Suppose that I accept A as well. The definition of $A \to B$ tells me that I am then forced to accept B. That's how you **use** my acceptance: $A \to B$, also A, therefore B (the elim rule in Table 3.4).

I can be persuaded to accept $A \to B$ if you can show me that whenever I accept A I must also accept B. This requires a hypothetical proof: ask me to suppose for the time being that I accept A, and show me that it would be

Table 3.4 Rules for implication

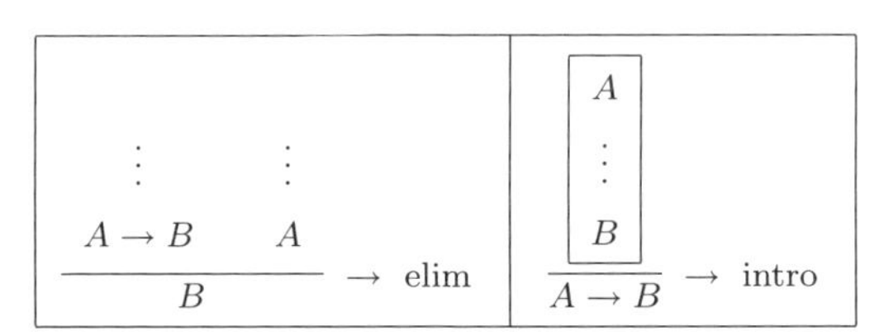

Suppose tomatoes cost £50 a kilo. Your name would be Richard then, wouldn't it?
So you must accept (tomatoes cost £50 a kilo) $\rightarrow$ (your name is Richard).

Fig. 3.1 The price of tomatoes

logically necessary, in that supposed situation, to accept B. Since I'm a fair-minded person, I must now accept that if ever I really did accept A, I would then also accept B, and that is the same as accepting $A \rightarrow B$. That how you **persuade** me to accept $A \rightarrow B$: if A then B, therefore $A \rightarrow B$ (the intro rule in Table 3.4).

All very well so far, but the definition of $A \rightarrow B$ means just what it says and no more. That economy has some famous consequences, some of which are disturbing.

3.2.2 One-way implication. If you are outside, then when it rains you get wet: *rain* $\rightarrow$ *wetting*. But if you are outside and you're wet, it might not be because of the rain: you might have been in the pool, or your best friend might have poured a bucket of water on your head.

The implication arrow is directional: it leads **from** A **to** B. If you have $A \rightarrow B$ you won't see A without B, says the definition — but it doesn't rule out B without A. Although A leads to B, it needn't be the only way to get there.

Reading the implication arrow backwards, seeing causes where there may be none, is called **abduction**. Beware of it in logical arguments.

Abduction seems to be essential if you are to guess truths from insufficient evidence. That is, it's necessary for everyday life, for science and for invention. It's an inspirational step, inductive rather than deductive, so not logical in our restricted sense of the word. A necessary step of reasoning which is outside logic? No wonder we can be so easily led astray — see the case of the drowned Major in Chapter 5.

3.2.3 Irrelevant implication: the price of tomatoes. Consider the argument in Fig. 3.1. It's disturbing. I have accepted for almost all my life that my name is Richard. I'm sure that my name has nothing to do with the price of tomatoes. But here is a hypothetical proof which seems to show logically that it does! I don't like it: the price of tomatoes has no effect on my name, it's strictly **irrelevant**.

Irrelevant implications, like the one in the price-of-tomatoes argument, are a consequence of the monotonicity principle. If I accept B, then when I add acceptance of A, by monotonicity I must still accept B. In any situation where I do accept A I will already accept B, so I can't evade the conclusion that I really ought to accept $A \to B$.

Irrelevant implications can't easily be eliminated from logical reasoning and therefore we shouldn't read the arrow too enthusiastically. All $A \to B$ tells us is "when we see A, then we see B". That sort of association wouldn't convince a wise court or a careful scientist that A causes B. When we have a cause, then we certainly have $\to$; when we have $\to$, we may or may not have found a causal relationship (see the cars and congestion argument in Chapter 5).

3.2.4 Useless implication: the cunning uncle. Suppose that you have a rich uncle who makes you a promise:

> *"Every year on your birthday, I shall give you £100."*

The uncle is as good as his word, and each year on your birthday you get the money. Not a fortune by today's standards, but not to be sniffed at: you can buy a lot of ice-creams with £100. An uncle who keeps his promises is surely a kind uncle (similar remarks apply, of course, to aunts and to large donations in any currency).

An unkind uncle might casually make the same promise, and then break it: your birthday comes around, but no money. You aren't any worse off than you would have been without the broken promise, but you feel as if you are. Unkind uncles are surely the worst.

Or are they? A cunning uncle might say to a young child

> *"Every 31st April, I shall give you £1,000."*

But there never will be such a day, so the cunning uncle never has to pay out. Is he kind or unkind: that is, has he broken or kept his promise? Well, he hasn't actually **broken** it: for him to be able to do that, there would have to be a 31st April one year, and on that day he would have to fail to pay out, and that day never happens. Logic takes a very literal legalistic view of these things, and it says that a promise is kept until it is broken. The cunning uncle gets away with it, in logical terms.

There's a sense — Chapters 9 and 10 make it precise — in which $A \to B$ is just a promise: when you see A, I promise that you will see B. If, like the cunning uncle's impossible payout day, I can never see A, the promise is kept — because it's impossible to break! The situation which disproves the $A \to B$ claim can't arise, so the claim is valid. On the other hand the circumstances in which we could use $A \to B$ can never arise, so the implication is strictly **useless**. Logicians call it **vacuous** implication.

It's not easy to build a logic without useless/vacuous implications: they arise from the consistency principle and the treatment of contradiction (Section 3.6). They're another reason why we should not read the arrow too enthusiastically: the definition means just what it says, and nothing more or less. If you never see A, then when you see A you see B.

3.3 Disjunction

Disjunction deals with situations in which there is more than one claim which you can accept. The formula $A \vee B$ is pronounced "A or B".

Definition 3.3 *When you accept $A \vee B$, you are forced to accept at least one of A and B.*

The point of this definition is its uncertainty. It doesn't say you're forced to accept A, and it doesn't say you're forced to accept B. It doesn't say you're forced to reject either of them. So you might accept just A, or you might accept just B, or you might accept both. The only certainty is that you're not allowed to reject them both.

Disjunctive uncertainty is expressed by the word 'or' in English. "We might go swimming or we might go to the cinema", for example.

> A subtlety which often trips up novices is that English 'or' usually means 'either/or'. To capture logical disjunction in English we usually have to say "A or B or both".

Multiple disjunctions bracket to the left: $A \vee B \vee C$ is read as if you'd written $(A \vee B) \vee C$. That means, according to the definition, 'you are forced to accept one of $A \vee B$ and C'; then because of the definition of $A \vee B$, that comes down, as it should, to 'you are forced to accept one of A, B and C'. But just as with conjunction, brackets affect form, not meaning, and you can prove the equivalence of $(A \vee B) \vee C$ and $A \vee (B \vee C)$.

3.3.1 Reasoning with disjunction. Suppose I accept A. Then, as a fair-minded person, I really ought to accept $A \vee B$, because I already accept one of A and B. That's a persuasion argument: A, therefore $A \vee B$. It doesn't matter how absurd or irrelevant B is, because $A \vee B$ doesn't claim that I accept B: A, therefore $A \vee B$, always. It's exactly the same the other way round: B, therefore $A \vee B$. There's no uncertainty in the arguments — uncertainty comes in when I tell you that I accept $A \vee B$ but I don't tell you how I was persuaded.

Suppose I accept $A \vee B$, and I refuse to tell you why. You know, because you know what $A \vee B$ means, that there are three possibilities — or, as philosophers call them, **cases**. Either

1. I accept A, or
2. I accept B, or
3. I accept both.

There is no hope of tricking me into revealing which of the three possibilities **is the case**, but there is something you can do: you can argue **by cases**, picking off the alternatives one by one, showing that each leads to the same conclusion so that the uncertainty is resolved.

You persuade me first that C is a logical consequence of A, picking whatever C suits your purposes. You persuade me next that the same C is also a consequence of B. Then you have persuaded me that I must accept C in any case: in case I accept A, then I must accept C; in case I accept B then I must accept C; in case I accept both, then by the monotonicity principle either the A-argument or the B-argument already shows that I accept C. So whichever alternative **is the case**, when I accept $A \vee B$ I must logically also accept C — that is, C is a logical consequence of $A \vee B$.

Argument by cases neatly sidesteps the uncertainty hiding in the claim $A \vee B$, never forcing me to reveal any bias I might have, but forcing me nevertheless towards a logical conclusion.

The formal rules for disjunction are in Table 3.5. The intro rules are as straightforward as the persuasion arguments (notice that they look like upside-down versions of the $\wedge$ elim rules). The elim rule, which captures the use argument, looks fearsome but (check it!) it really is no more than the argument by cases.

3.3.2 This is *un*certainty? It certainly doesn't feel like it to me! The elimination (use) rule is fine: it neatly acknowledges and sidesteps the uncertainty implicit in $A \vee B$. The introduction rules, on the other, use certainty and then hide it. A particular kind of uncertainty, perhaps, but nothing which captures the idea that $A \vee B$ really ought to be established in a situation where I'm sure

Table 3.5 Rules for disjunction

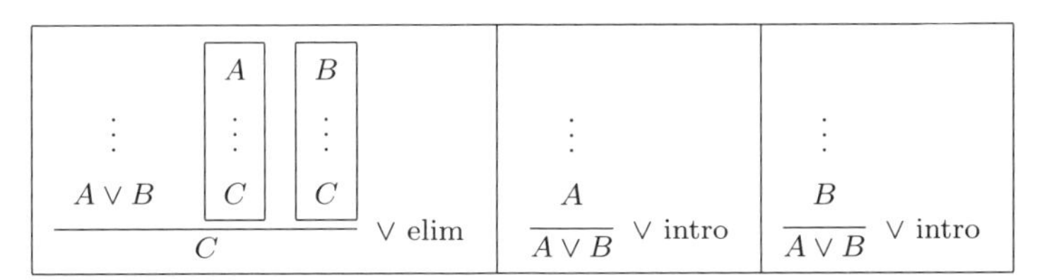

I accept one or more, but I'm not sure which. That's trickier than it sounds: the problem will turn up again in the treatment of $\exists$ in Chapter 6 and in the discussion of the universal drunk in Chapters 7 and 10.

3.4 Negation and contradiction

Negation deals with the possibility of a claim being wrong. The formula $\neg A$ is pronounced "not A". The claim it makes is something like "when you accept $\neg A$, you would be wrong to accept A". Since the only way of being wrong in logic is to accept a contradiction, $\neg A$ must mean

Definition 3.4 *When you accept $\neg A$, accepting A leads to a contradiction.*

Multiple negations are perfectly ok: you can write $\neg\neg\neg\neg A$, and it is read, as you would expect, as if you'd written $\neg(\neg(\neg(\neg A)))$.

According to the definition, $\neg\neg A$ means "(A leads to a contradiction) leads to a contradiction".

Don't assume that $\neg$ is like numerical negation, so that $\neg\neg A$ is automatically equivalent to A. This is formal logic, not Boolean-algebra arithmetic.

3.4.1 Reasoning with negation. The meaning of negation is tangled up with the meaning of contradiction, and to show how we deal formally with $\neg A$ and $\neg\neg A$ and so on, it's necessary to talk about contradiction.

The contradiction symbol ($\bot$) is not a connective, it's a formula. It describes impossibility, confusion, a situation that can't happen. The symbol is called 'bottom' by mathematically inclined computer scientists.

To persuade me to accept $\neg A$, show me that if I accept A, then I would have to accept a contradiction (the intro rule in Table 3.6). To persuade me that I already accept a contradiction, show me that I accept some claim and at the same time it would be wrong to accept it (the elim rule in Table 3.6).

Table 3.6 Rules for negation

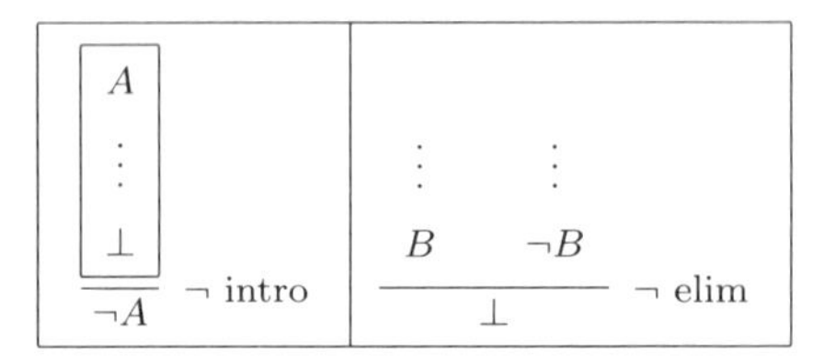

If negation depends on contradiction, what does contradiction mean? To answer that question it's necessary to take a well-trodden philosophical diversion.

3.5 Is there a 'law of excluded middle'?

At this point just about everything that's odd about formal logic is ready to be exposed, because it's nearly all bound up with negation and contradiction. The peculiarities of disjunction and implication (apart from irrelevant/price-of-tomatoes implications, which are caused by monotonicity) come back to negation and contradiction and how we use them.

It turns out that there are two ways to proceed from here. Technically it can all be boiled down to a single choice of rule; philosophically the choice is complicated and important.

Frege, remember, started his work by asking a philosophical question: "how do we know the truths of arithmetic?". For us the question is "what do logical formulae claim?". In this book, because I'm interested in making proofs, I've taken the position that a formula A claims "I have a proof of A". That makes particular sense in the world of computer programming and computer science. Most mathematicians outside computer science would take a different position, and say that A claims "A is true".

These two positions sound as if they are universes apart and incomparable. But we can compare them by asking a single technical question: "am I forced to accept $A \vee \neg A$?". Those who take the truth-claim position call $A \vee \neg A$ the **law of excluded middle**: either a formula is true or it's false, and there is no other possibility. Those who take the proof-claim position don't accept that excluded middle is a law. The battle is an old one, going back at least a century. You don't have to take a position, but you do need to understand the ground.

3.5.1 The constructive position. Constructivists — and this book smiles on the constructive position for the most part — only accept claims that have been proved. To make a constructivist accept $A \vee \neg A$, using the rules you've seen so far, you either have to show a proof of A, or you have to show a proof of $\neg A$. There are lots of As for which you can't do that.

Science contains lots of **undecided** questions: must the universe continue expanding? is there life on Mars? is there a cure for the common cold? As I write we can't prove that the answer to any of the questions is "yes" or that it is "no", even when we are pretty certain sure that in practice we know the answer.

Everyday life throws up less profound questions too, undecided within the present instant, but decided if you wait long enough, like "it will rain tomorrow". You can't prove it will rain — not even under a black cloud in the Lake District in the winter — and you can't prove it won't — not even in the Sahara desert in the dry season.[1]

The world of mathematics, you might imagine, has a firmer grasp on the issue. For centuries, Fermat's last theorem:

> *"To resolve a cube into the sum of two cubes, a fourth power into two fourth powers, or in general any power higher than the second into two of the same kind, is impossible; of which fact I have found a remarkable proof. The margin is too small to contain it."*

— often summarized as

> *there are no solutions of the equation* $a^n + b^n = c^n$ *for integers* $a,\ b,\ c > 0$ *and integer* $n > 2$

— was once everybody's favourite example of a mathematical statement which surely ought to be true, but which nobody could prove. Now that it seems at last to have been proved, we can fall back on Goldbach's conjecture

> *every even integer* $n > 2$ *is the sum of two primes*

— again, everybody thinks it's true, but nobody has yet proved it. (If Golbach[2] is ever proved right, there are plenty more unproved conjectures in the rich field of prime numbers: this chapter will only need minor changes.) $\mathit{Goldbach} \vee \neg\mathit{Goldbach}$? Who knows, say the constructivists — we neither have a proof, nor do we know that a proof is impossible.

The mathematical philosophy called **intuitionism** can be summarized as saying that mathematics is all made up by humans. Intuitionists hold that mathematical 'truth' is only what we've persuaded each other to accept by exchanging proofs, and they only accept constructive proofs. Intuitionism is often confused with constructivism but they aren't identical: you don't have to be an intuitionist to be a constructivist.

3.5.2 The classical position. Most mathematicians don't hold with constructivism, still less with intuitionism. Those who take the 'classical' or 'non-constructive' position accept the law of excluded middle, backed up by the philosophy called **Platonism** (named for the Ancient Greek philosopher Plato) which can be summarized as saying that mathematical truths are real and they

[1] So is it **wrong** to believe it will rain tomorrow? Not really: it's neither right nor wrong; it's a gamble. It's only wrong to believe that you can **prove** it will rain tomorrow.

[2] Goldbach said every $n > 5$ and three primes, supposing 1 to be a prime. Euler produced the version quoted here, but Euler wasn't short of fame so it's still called Goldbach's conjecture. Sometimes it's stated as $n > 4$ and two odd primes, but of course that's the same conjecture because $4 = 2 + 2$ and 2 is the only even prime.

exist whether we know them or not. Our job is to find out truth, not to invent it. From this point of view the statement "Goldbach's conjecture holds" is either certainly true or certainly false, even though neither you nor anybody else just now has any way of knowing which it is. Most working mathematicians are Platonists.

3.5.3 The difference, by argument. Consider the formula $(A \to B) \vee (B \to A)$. It's classically provable and constructively disprovable because it's a logical consequence of accepting the law of excluded middle. It seems, at first sight, to suggest that there's always a connection between two claims, whatever they are. To take an extreme example:

> *either love of fishing leads to hatred of bicycles,*
> *or hatred of bicyles leads to love of fishing*

Surely that's absurd! says the constructivist.

No, says the classicist, it's not absurd at all. Either you love something or you hate it **already**, whether you know it or not. The only way to deny $A \to B$ is to accept A and deny B, so if you deny that your love of fishing leads you to hate bicycles, you must love fishing and not hate bicycles. But then you love fishing, so by the price-of-tomatoes trick (ouch!), in your case hatred of bicycles would lead to love of fishing. And the other way round if you try to deny the second implication.

Oh yes it is absurd, says the constructivist. $A \to B$ means 'now, and for always, and whatever happens, when I accept A I will at the same time accept B'. My feelings about fishing and bicycles aren't sorted out yet — I'm still making my mind up. That doesn't mean I won't ever make my mind up (that would mean $\neg A$ and $\neg B$, and you could play the useless implication trick). But you can't say $A \to B$ because I may find out, after a couple of cold days in a boat, that I love fishing, but I might never get on a bicycle and learn to hate it. And you can't say $B \to A$ for the same reason. Sure, in the future one or other or both those claims might be valid, but they aren't valid **now**.

We can leave them squabbling: each, on their own ground, is certain that they are right.

3.5.4 The difference, by mathematical example. If you find arguments about fish and bicycles a little too informal, Fig. 3.2 shows a famous example of a mathematical proof that classicists accept but constructivists reject. To understand the claim you should recall that a **ratio**nal number is one that can be written as the **ratio** of two integers — i.e. as a fraction — and an ir**ratio**nal number, like π or e or $\sqrt{2}$, is one that can't be.

Claim: There are irrational numbers x and y such that x^y is rational.

Proof: Premise: $(\sqrt{2})^{\sqrt{2}}$ is either rational or irrational.
Case 1: $(\sqrt{2})^{\sqrt{2}}$ is rational.
Put $x = y = \sqrt{2}$, and we have found an example.
Case 2: $(\sqrt{2})^{\sqrt{2}}$ is irrational.
Put $x = (\sqrt{2})^{\sqrt{2}}$, $y = \sqrt{2}$;
$x^y = ((\sqrt{2})^{\sqrt{2}})^{\sqrt{2}} = (\sqrt{2})^{(\sqrt{2}\times\sqrt{2})} = (\sqrt{2})^2 = 2$;
2 is rational (it's $\frac{2}{1}$), and we have found an example.

Fig. 3.2 A classical proof, unacceptable to constructivists

Classicists accept the premise of the proof, because of the law of excluded middle, and the rest follows (an argument by cases, in effect an instance of $\vee$ elim).

Constructivists say "show me this number $\sqrt{2}^{\sqrt{2}}$ and demonstrate whether it is rational or not!", and since we don't know how to do that (we really don't!) they need not accept the premise, and therefore the proof as a whole.

3.5.5 The consequence. If you take the constructivist position then, as you will see when you experiment with Jape, $\neg\neg A$ isn't identical to A, $A \vee \neg A$ doesn't always hold, $(A \rightarrow B) \vee (B \rightarrow A)$ is similarly doubtful, and so on and on. Jape has a panel labelled 'Classical conjectures' of claims which hold in classical logic, but not constructively.

On the other hand, classicists accept everything that constructivists do, and more besides.

The overall situation is summarized in Fig. 3.3. Everything that can be claimed is inside the box. Everything in the oval is classically provable; nothing outside it is. Everything in the blob is constructively provable; nothing outside it is. Since the blob is part of the oval, there's a disputed region: some claims are classically but not constructively provable. Conjectures outside the oval — most conjectures, in fact — are neither classically nor constructively provable. Conjectures inside the blob are both classically and constructively provable.

You can look at this situation in two ways. From the classical point of view, constructivists are too fussy. There are proofs in the grey region which classicists can easily make, but constructivists just refuse to accept. So, you might say, classical logic is just more **powerful**, more **useful**. On the other hand, constructivists have less work to do: the constructive rule for contradiction mean they have to prove fewer theorems than the classicists do, and there are classical proofs which seem to support absurd claims, as we shall see. So, you might say, constructive logic is **safer**.

3.5.6 Which way to go? There is a philosophical debate about what mathematics really means, and no important philosophical question ever gets a final answer. This one will run and run. If you want to join in you can, or you can just let them get on with it. The good news is that as a **user** of logic

you don't have to take sides!

You ought to know, as a user, that there are different logics, because they have different characteristics and different uses. Logic, for the practising computer scientist, is a playground: we can play with any ball that we can pick up. This book deals with three logics (constructive and classical Natural deduction in Parts II and III, Hoare logic for programs in Part IV), but there are lots more out there. You pick one to suit your purposes. Constructive logic has a particular link to formal calculation, and that makes it relevant to reasoning about declarative programs. Classical logic is usually used to reason about imperative programs (see Part IV). Horses for courses; different strokes for different folks; it's all logic.

The presentation in the first two parts of this book leans towards constructive logic, but that's for pragmatic reasons. Since constructive formal proof is easier than classical formal proof (it really is!) and since Jape makes that fact clear to anybody who plays with it, I've given definitions of the connectives in constructive style. Then, despite the fact that constructive **disproof** is a good deal trickier than classical, I've had to be honest and show the underpinnings of constructive logic in Part III. Finally, in Part IV, classical logic takes over in Hoare logic's treatment of imperative programs.

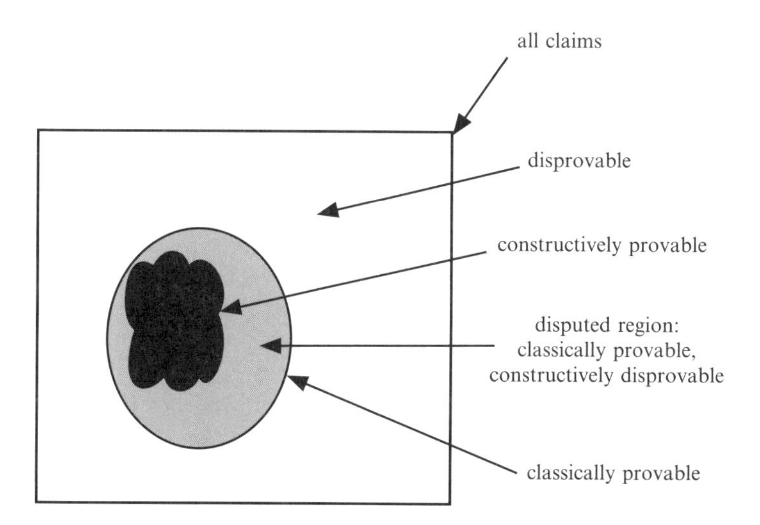

Fig. 3.3 Proofs, disproofs, and disputed conjectures

I quite like the ideas of constructivism, but that's just me. To read this book you really **don't** have to make a commitment. Jape includes both constructive and classical versions of the contradiction rule. It lets you swing both ways and build classical proofs and constructive disproofs (see Chapter 9), both at the same time, of claims that fall in the disputed region of Fig. 3.3. My aim is to let you know that there's more than one logic out there, to show you how subtle the differences between logics can be, and then to turn you loose in the logical playground.

3.6 Rules for contradiction

Contradiction is a formula. It's a formula which represents confusion. What can you do with confusion? Surprisingly, quite a bit.

3.6.1 Constructive reasoning with contradiction. Recall the consistency principle: contradictions can't be allowed. If a contradiction arises in a proof, we are in an impossible situation. In impossible situations, it doesn't matter what we do — because impossible situations can't happen.

Consider how we should reason if we accept $A \vee B$ and at the same time $\neg B$. We know that we accept either A or B or both. But we also know that it would be wrong to accept B at all, because we already accept $\neg B$. So the only reasonable possibility is to accept A. This happens quite a bit when reasoning by cases: one of the cases is impossible, and need not be taken seriously.

When we translate this into a formal disjunction step, we seem to be stuck:

$$\dfrac{\begin{array}{c}\vdots\\ A \vee B\end{array} \quad \boxed{\begin{array}{c}A\\ \vdots\\ A\end{array}} \quad \boxed{\begin{array}{c}B\\ ?\\ A\end{array}}}{A}\ \vee \text{ elim} \tag{3.1}$$

It's obvious that if we accept A we accept A, so the first case can be dealt with. But how does A follow from B? What follows from B, because we have the premise $\neg B$, is a contradiction, which according to the consistency principle means that the second case is impossible, and ought to be ignored.

The technical way in which Natural Deduction deals with this situation is to say that in an impossible situation, you can conclude whatever you like in order to tidy things up. It doesn't matter what you do, because it doesn't matter: that situation will never happen. In short, if you meet a contradiction, write the subproof off. In an impossible situation, conclude anything you like (and don't bother me again!). That's the constructive contra rule in Table 3.7. Both constructivists and classicists accept it, and both can use it.

Table 3.7 Rules for contradiction

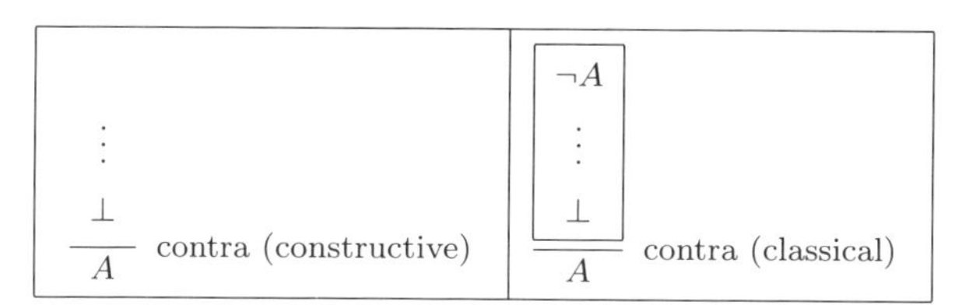

We can derive any conclusion from a contradiction, then. If we have to show A from B and $\neg B$, we don't really have to worry. A situation in which both B and $\neg B$ hold will never arise, so we can conclude A if we need to.

$$\dfrac{\begin{array}{c}\vdots\\ A \vee B\end{array} \quad \boxed{A} \quad \boxed{\begin{array}{c}B\\ \neg B\\ \bot\\ A\end{array}}}{A}\ \vee\ \text{elim} \tag{3.2}$$

(This diagram only suggests a proof: the proof proper is shown on page 56 in Chapter 5.)

What is remarkable is that a technical fix, essential to deal with can't-happen cases, fits so well with the rest of the logical machinery, and doesn't lead into paradox. It's all that constructivists need to know.

3.6.2 Classical reasoning with contradiction. Classicists accept the constructivist argument about use of contradiction, but they don't think it goes far enough. They accept, remember, that either A is true (whether we can prove it or not) or else $\neg A$ is true (likewise). For them $\neg A$ is just the opposite of A.

The $\neg$ intro rule lets us derive $\neg A$ if A leads to a contradiction. Classicists also do it the other way round, in the classical contra rule of Table 3.7. This argument-shape is what's called **proof by contradiction**. To show that A must hold, suppose that it doesn't and show that then there is a contradiction. That proves it's impossible that A doesn't hold — so, for a Platonist, it's certain that it **does** hold.

3.6.3 The classical rule includes the constructive. Classical proof-by-contradiction introduces the assumption $\neg A$ into the argument. Assumptions are resources you can use to reach a conclusion. Proving a contradiction from the current premises and assumptions is the constructivists' task; the classicists get extra help. That is, there will be circumstances in which the classical rule works, because of the extra assumption, but the constructive rule doesn't. That's either an advantage (if you're looking for **some way** to conclude A) or a disadvantage (if you are doubtful that you really want to believe in A).

By the monotonicity principle, if you can prove a contradiction from the premises, you can prove it from the premises plus $\neg A$. So, given what you know about logic already, you should be able to understand that:

> If you accept the classical treatment of contradiction, you accept the constructive treatment as well — but not vice-versa.

That means that you can imitate any constructive contra step in classical logic. Here, for example, is a sketch of the proof that from $A \vee B$ and $\neg B$ you can conclude A, using the classical contra rule (the formal proof, is shown on page 56 in Chapter 5):

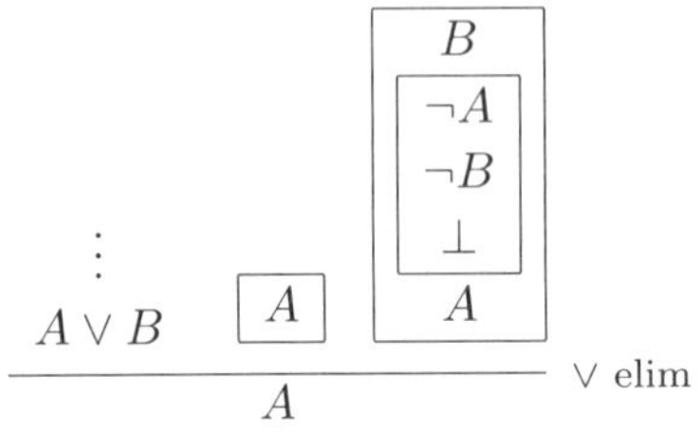

$$\dfrac{\begin{array}{c}\vdots\\ A\vee B\end{array}\quad \boxed{A}\quad \boxed{\begin{array}{c}B\\ \boxed{\begin{array}{c}\neg A\\ \neg B\\ \bot\end{array}}\\ A\end{array}}}{A}\ \vee\text{ elim} \qquad (3.3)$$

The contradiction still comes from the B assumption and $\neg B$ which we assumed already; the extra assumption $\neg A$ is unnecessary, but by the monotonicity principle it doesn't get in the way.

3.7 Truth is trivial

After all that fuss about contradiction, it is a relief to find that the treatment of truth in Natural Deduction is really simple. You can always conclude $\top$ (Table 3.8). That's it: you don't need any premises, any antecedents, any support.

Table 3.8 The truth rule

$$\boxed{\dfrac{}{\top}\ \text{truth}}$$

Unfortunately that really **is** it: there is **absolutely nothing** more to say about $\top$. It's a shame to discover that — because there is no other rule which involves the symbol — you can't use a proof of $\top$ to prove anything (if you could, you could prove every formula, because $\top$ is always available — that's why it has to be useless).

Constructivists, who deal in proof, don't have to think of $\top$ as 'truth'. They see it as a counterpart of $\bot$ (but they still find it useless).

Table 3.9 Rules for connectives summarized

Introduction (persuasion)	Elimination (use)
$\dfrac{\begin{matrix}\vdots & \vdots \\ A & B\end{matrix}}{A \wedge B}$ $\wedge$ intro	$\dfrac{\begin{matrix}\vdots \\ A \wedge B\end{matrix}}{A}$ $\wedge$ elim $\quad$ $\dfrac{\begin{matrix}\vdots \\ A \wedge B\end{matrix}}{B}$ $\wedge$ elim
$\dfrac{\boxed{\begin{matrix}A \\ \vdots \\ B\end{matrix}}}{A \to B}$ $\to$ intro	$\dfrac{\begin{matrix}\vdots & \vdots \\ A \to B & A\end{matrix}}{B}$ $\to$ elim
$\dfrac{\begin{matrix}\vdots \\ A\end{matrix}}{A \vee B}$ $\vee$ intro $\quad$ $\dfrac{\begin{matrix}\vdots \\ B\end{matrix}}{A \vee B}$ $\vee$ intro	$\dfrac{\begin{matrix}\vdots & \boxed{\begin{matrix}A \\ \vdots \\ C\end{matrix}} & \boxed{\begin{matrix}B \\ \vdots \\ C\end{matrix}} \\ A \vee B & & \end{matrix}}{C}$ $\vee$ elim
$\dfrac{\boxed{\begin{matrix}A \\ \vdots \\ \bot\end{matrix}}}{\neg A}$ $\neg$ intro	$\dfrac{\begin{matrix}\vdots & \vdots \\ A & \neg A\end{matrix}}{\bot}$ $\neg$ elim

Table 3.10 Rules for truth and contradiction symbols summarized

$\dfrac{}{\top}$ truth	
$\dfrac{\begin{matrix}\vdots \\ \bot\end{matrix}}{A}$ contradiction (constructive)	$\dfrac{\boxed{\begin{matrix}\neg A \\ \vdots \\ \bot\end{matrix}}}{A}$ contradiction (classical)

3.8 The logical connectives summarized

Formal logic really is simple. It needs only ten rules for the logical connectives — Table 3.9 — plus one each for truth and contradiction — Table 3.10. Contradiction comes in two forms; the classical includes the constructive.

The simplicity is real: those rules, plus four more for the quantifiers (see Table 6.5 on page 95) are all there is to formal proof. Of course you have to learn how to use the rules ... which is what the next two chapters are about.

4 Rule shapes and formula shapes

The inference rules of Natural Deduction are patterns, shapes, **schemes** which you can use to make particular proof steps. You use a rule scheme to make an **instance** that suits your needs. Verbing the word 'instance', we describe the activity as **instantiating** the rule.

The names A, B and C are **parameters** of the rule schemes of Chapter 3. You make an instance by replacing each of the parameters of a scheme by a formula that you choose. Of course you must replace every occurrence of A by a copy of the A-formula, and likewise the Bs must all be replaced by copies of the B-formula, and so on, but otherwise you have complete freedom: you can use the same formula for A as for B, or a different one, just as you wish.

The formulae that replace the parameters to make an instance are called **arguments**. That would be more than a little confusing in a book which is about proof and which equates 'argument' with 'piece of reasoning'. But I have to call them something, and I don't want to invent a completely new name, so in this chapter I've called them **formula arguments**. Formula arguments replace scheme parameters to make instances.

In this book I use A, B and C as parameter names but E, F, G and H in formula arguments, to make a clear separation between schemes and instances. So to make an instance of the $\wedge$ intro rule, for example, I can put $E \rightarrow F$ in place of each occurrence of A, and $F \rightarrow E$ in place of each occurrence of B. To make sure I get the right formula structure and to make it clear what I'm doing in this example, I bracket each copy of each formula argument: see Fig. 4.1. Everything that Chapter 3 says about persuasion and use — persuade me to accept A and separately to accept B, and I can't refuse to accept $A \wedge B$; persuade me to accept $A \wedge B$ and I must accept A and also B — applies if you read $E \rightarrow F$ in place of A, and $F \rightarrow E$ in place of B. Or indeed any two formulae at all: contradictory, complementary, identical, it's all the same. Fig. 4.2 shows some more instances of the same rule-schema, with brackets round non-atomic insertions.

$$\dfrac{\begin{matrix}\vdots \\ (E \rightarrow F)\end{matrix} \qquad \begin{matrix}\vdots \\ (F \rightarrow E)\end{matrix}}{(E \rightarrow F) \wedge (F \rightarrow E)}\ \wedge\ \text{intro}$$

Fig. 4.1 An instance of $\wedge$ intro

$$\dfrac{\begin{matrix}\vdots & \vdots \\ E & (\neg E)\end{matrix}}{E \wedge (\neg E)} \qquad \dfrac{\begin{matrix}\vdots & \vdots \\ (F \vee G) & (G \wedge F)\end{matrix}}{(F \vee G) \wedge (G \wedge F)} \qquad \dfrac{\begin{matrix}\vdots & \vdots \\ E & E\end{matrix}}{E \wedge E}$$

Fig. 4.2 Three instances of $\wedge$ intro

4.1 What counts as a formula?

E, F, G and H are each a formula; so are $\bot$ and $\top$. You can make bigger formulae out of smaller by using connectives: $(A \wedge B)$, $(A \to B)$, $(A \vee B)$, $(\neg A)$ are schemes for making larger formulae out of smaller ones. So, for example, $(\neg E)$ is a formula, and therefore $((\neg E) \to F)$ is a formula. So is $(G \vee \bot)$, and therefore $(((\neg E) \to F) \wedge (G \vee \bot))$ is a formula. And so on — we can build up formulae of any size we like using whatever symbols and connectives we like.

But it seems we must use rather a lot of brackets! I'll deal with that problem later.

4.1.1 How to pronounce brackets. Mostly we don't pronounce brackets: $(A \wedge B) \wedge C$ is usually read as "A and B" (pause) "and C". But you have to say something special to pronounce $\neg(\neg(\neg(\neg C)))$.

It's ok to say 'left bracket' and 'right bracket'; it's ok to say 'open bracket' and 'close bracket'; it's ok to say anything so long as the people who are listening know what you're talking about. Some people, picking up on Dirac's quantum-mechanical notation, say **bra** and **ket**. (Sniggering is not allowed, and those who have been waiting for an opportunity to insist that we should say 'parenthesis' instead of bracket can crawl right back into their holes right now.)

4.2 Shape matching: fitting a formula to a scheme

To make proofs you have to be able to look at a formula and decide which rules you might apply to it. It's a matter of matching shapes mechanically, and it has nothing to do with what the formula means. Formulae **fit** schemes; schemes **match** formulae.

If a formula is fully bracketed — a pair of brackets for each connective — then it's easy to see how to take it apart, and that shows what kind of formula it is. For example, $(((\neg E) \to F) \wedge (G \vee H))$ is definitely a conjunction. It fits the $(A \wedge B)$ scheme, and the brackets allow only one reading:

$$\underbrace{(\underbrace{(\underbrace{(\neg E)}_{(\neg A'')} \to F)}_{(A' \to B')} \wedge \underbrace{(G \vee H)}_{(A''' \vee B''')})}_{(A \wedge B)}$$

It doesn't fit any of the other formula schemes. $(A \vee B)$ won't match it, for example, even though it contains a $\vee$ connective: you would have to match A with "$((\neg E) \rightarrow F) \wedge (G$" and B with "$H)$", and in neither case have you matched a parameter to a properly made formula, because the brackets don't balance.

Once you know what formula scheme a formula fits, you immediately know what rule schemes it fits. A conjunction, for example, fits the conjunction rules. $(((\neg E) \rightarrow F) \wedge (G \vee H))$ fits $(A \wedge B)$ as above — matching A with $((\neg E) \rightarrow F)$ and B with $(G \vee H)$ — and that tells us how to instantiate each of the conjunction rule schemes, giving three possible rule-instances and therefore three possible proof steps.

$$\dfrac{\begin{matrix}\vdots & & \vdots \\ ((\neg E) \rightarrow F) & & (G \vee H)\end{matrix}}{(((\neg E) \rightarrow F) \wedge (G \vee H))}\ \wedge \text{ intro}$$

$$\dfrac{\begin{matrix}\vdots \\ (((\neg E) \rightarrow F) \wedge (G \vee H))\end{matrix}}{((\neg E) \rightarrow F)}\ \wedge \text{ elim} \qquad \dfrac{\begin{matrix}\vdots \\ (((\neg E) \rightarrow F) \wedge (G \vee H))\end{matrix}}{(G \vee H)}\ \wedge \text{ elim}$$

The message is that the **shape** of a formula shows you what rules you can use to work on it. **Shape-matching** is the basis of proof search. Since our business is proof, we have to understand formula shapes, and that's what this chapter is really about.

4.3 Determining the shape of an unbracketed formula

Formal reasoning is reasoning by shape. Brackets in a formula perfectly delineate its shape, but brackets are serious visual clutter. They make the formula harder to read, even though they make it easier to explain what it means. In practice we reduce clutter by obeying a convention that determines the shape of an unbracketed formula, leaving brackets to be used for emphasis or in exceptional situations when we don't want the convention to apply. In essence, the convention is a means of telling you where the brackets would have to go if you bothered to put them in.

The convention I use in this book is quite standard, and is based on giving each connective a **binding priority** and a **binding direction**. It's the mechanism that Jape uses to recognize the shape of formulae, so it's worth understanding it and getting used to it.

4.3.1 Calculating the value of an arithmetic formula. We learn to read arithmetic formulae in first school, but we are taught to read them as calculations. We learn calculation slogans: I was taught the ones in Table 4.1.

Table 4.1 Slogans for arithmetic calculation

• work inside brackets first!
• negations next!
• multiplication and division before addition and subtraction!
• work left to right!

The list of slogans lengthens, and calculation gets more intricate, when you learn that you can do additions and multiplications in any old order — e.g. $(5+2)+1$ gives the same result as $5+(2+1)$ — but divisions and subtractions are strictly left-to-right. It gets more intricate still when you learn syntactic equivalences which convert additions into subtractions — e.g. $5-(2+1)$ gives the same result as $5-2-1$ — divisions into multiplications, erase double negations, and so on.

Because of the left-to-right slogan, children read $5 - 2 - 1$ as "start with 5; subtract 2; subtract 1", and thus work out that the result is 2. If we were to insert brackets to imitate the left-to-right reading — one pair of brackets per operator — we would write $((5 - 2) - 1)$. And that's the only way to do it: right-to-left bracketing $(5 - (2 - 1))$ gives the wrong answer — 4 instead of 2.

$5 + 2 - 1$ is trickier: the left-to-right reading "5; add 2; subtract 1" — i.e. $((5+2)-1)$ — gets the right answer, of course, but so does "2; subtract 1; add 5" — i.e. $((2-1)+5)$ which, because $A+B = B+A$, is equivalent to $(5+(2-1))$. Clearly, school-style calculation isn't entirely straightforward.

When we have mixed operators the slogans tell us which way to work. Because we do multiplications before additions, $4 + 6 \times 7$ is always "take 6; multiply by 7; add 4" — i.e. $((6 \times 7) + 4)$ which is equivalent to $(4 + (6 \times 7))$. To get the $((4 + 6) \times 7)$ reading you have to put in some brackets first.

Negation is treated as a device for changing the sign of a number: $-4+6$, for example, is seen as an addition of the numbers -4 and 6; $-(4+6)$ calculates 10 and then changes it into -10. If you were allowed to write $--4+6$ the negations would first change 4 into -4 and then -4 back to 4, so double negations clearly cancel.

4.3.2 Determining the shape of an arithmetic formula. All of that calculation-specific knowledge, hammered in so hard and so deep when you were so very young, has a corresponding treatment in terms of formula shapes. Indeed, it's almost the same treatment. The main difference is the use we make of the shapes: when **classifying** a formula we work from the outside inwards; when **calculating** we work from the inside outwards.

We determine the shape of a formula by putting brackets round its operators in order of **binding power** — roughly, the syntactic 'strength' of the operator. We have to know how to treat apparently overlapping occurrences of similar-strength operations, so we give a **binding direction**: right-to-left for negations, so $--4$ is $(-(-4))$; left-to-right for the others, so $4\times7\div3\times6$ is $(((4\times7)\div3)\times6)$. Table 4.2 gives the list, with the strongest-binding operator first.

Table 4.2 Binding power and binding direction of arithmetic operators

1. negation (right-to-left)
2. multiplication, division (left-to-right)
3. addition, subtraction (left-to-right)

The underlying principle is the same as the calculation slogans. Instead of "calculate negations first", we get "bracket negations first", and so on: no big deal. There are some surprises, though: we don't consider syntactic equivalences — no converting $6 + (-4)$ into $6 - 4$, for example — and the bracketing rules allow some formulae which first-schoolers might be surprised at, like $64/2/4$, equivalent to $((64/2)/4)$, or 7×-4, equivalent to $(7 \times (-4))$. Together with the principle that you don't invade bracketed components — $(4 + 6) \times 7$ is 10×7, not $4 + 42$, despite the relative strengths of the operators — the table is all you need.

Consider $3 \times 2 - 4 + 6 \times 12$, for example. Bracket negations first (none in this example, so nothing to do). Next multiplications and divisions: there are two multiplications but they don't overlap, so the binding power doesn't matter and we bracket them separately, giving $(3 \times 2) - 4 + (6 \times 12)$. Last, additions and subtractions: there is one of each, and they do overlap, so binding power comes into play. We must bracket left-to-right, taking the subtraction first, giving $((3 \times 2) - 4) + (6 \times 12)$; last of all we bracket the addition, giving $(((3 \times 2) - 4) + (6 \times 12))$.

I can't remember whether as a child I ever met a formula like $---17 \times 3 \times -(4 + -6)$, but in any case our rules can deal with it. There are five negations, three of them overlapping which have to be bracketed right-to-left, giving $(-(-(-17))) \times 3 \times (-(4+(-6)))$. Then all that's left unbracketed is a couple of multiplications, which are bracketed left-to-right to give $(((-(-(-17))) \times 3) \times (-(4 + (-6))))$.

Exercise 4.1 Bracket each of the following formulae, using the priority scheme of Table 4.2.

1. $27 + 3$
2. $27 - 5 + 6$
3. $27 + 5 - 6$
4. $27 \times 5 + 6$
5. $27 - 5 \div 6$
6. $(27 \quad 5) \times 6$

7. -4
8. -4×2
9. -4×-2
10. $-(4 \times -2)$
11. $(-4) \times -2$
12. $(-4) \times (-2)$
13. $-27 + 6 \times 5 \div 32 - -6$

4.3.3 Finding the principal operator. $5 - 2 - 1$ is a subtraction because the outermost operator in the fully bracketed form $((5-2)-1)$ is a subtraction. That operator — the rightmost subtraction in the unbracketed form — is the **principal operator** of the formula and its **operands** are $5-2$ and 1. In $3 \times 2 - 4 + 6 \times 12$ the principal operator is the addition, because it brackets as $(((3 \times 2) - 4) + (6 \times 12))$, and its operands are $3 \times 2 - 4$ and 6×12. In $---17 \times 3 \times -(4+-6)$ it's the second multiplication, because it brackets as $(((-(-(-17))) \times 3) \times (-(4 + (-6))))$, and its operands are $---17 \times 3$ and $-(4 + -6)$. In $-(4 + -6)$ it's the negation and its operand is $(4 + -6)$.

Once you can spot a principal operator you can fit a formula to a formula scheme. If the principal operator is a subtraction then it fits $A - B$; if it's a multiplication then it fits $A \times B$, and so on. In turn, that tells you how to take the formula apart: if the principal operator is a subtraction then it fits $A - B$, A matches the stuff to the left of the subtraction (after you've stripped away any bracket-pairs that enclose the whole formula), and B matches the stuff to the right.

Exercise 4.2 In each of the following formulae, circle the principal operator, using the priorities and binding directions of Table 4.2. Avoid bracketing the formula if you can.

1. $E \times F - G \div H$
2. $E \times F - (G \div H)$
3. $E \times (F - G) \div H$
4. $-E \times (F - G \div H)$
5. $-(E \times F - G) \div H$
6. $-(E \times F - G \div H)$
7. $-E \times -F - G \div -H$
8. $-(E \div G) \times (H - E1\ ) \div -G1$

4.3.4 Finding the principal connective. The good news is that logical formulae can be treated exactly like arithmetic formulae. Table 4.3 gives the binding powers of the connectives. There are no surprises, apart from the fact that $\rightarrow$ binds right-to-left (i.e. $A \rightarrow B \rightarrow C$ brackets as $(A \rightarrow (B \rightarrow C))$).

The ordering of the connectives in the table is arbitrary and the binding directions, apart from negation which, as a prefix connective, necessarily binds right-to-left, are arbitrary too. But it's a pretty simple system, easy to learn and, because this book uses it and Jape uses it too, you have to learn it.

Table 4.3 Binding power and binding direction of logical connectives

1. $\neg$ (right-to-left)
2. $\wedge$ (left-to-right)
3. $\vee$ (left-to-right)
4. $\rightarrow$ (right-to-left)

Exercise 4.3 Bracket each of the following formulae, using the priorities and binding directions of Table 4.3.

1. $E \wedge (F \vee G \rightarrow H)$
2. $(\vee E \wedge F) \vee G \rightarrow H$
3. $\neg E \rightarrow F \rightarrow G \rightarrow H$
4. $(E \wedge F) \vee (G \wedge H)$
5. $E \vee F \wedge (G \rightarrow H)$
6. $E \wedge F \vee G \vee H$
7. $\neg E$
8. $\neg\neg E$
9. $\neg E \wedge \neg F$
10. $\neg(E \wedge \neg F)$
11. $(\neg E) \wedge \neg F$
12. $(\neg E) \wedge (\neg F)$
13. $\neg E \vee G \wedge H \rightarrow E1 \vee \neg G1$

Exercise 4.4 In each of the following formulae, circle the principal connective, using the priorities and binding directions of Table 4.3. Avoid bracketing the formula if you can.

1. $E \wedge F \rightarrow G \vee H$
2. $E \wedge F \rightarrow (G \vee H)$
3. $E \wedge (F \rightarrow G) \vee H$
4. $\neg E \wedge (F \rightarrow G \vee H)$
5. $\neg(E \wedge F \rightarrow G) \vee H$
6. $\neg(E \wedge F \rightarrow G \vee H)$
7. $\neg E \wedge \neg F \rightarrow G \vee \neg H$
8. $\neg(E \vee G) \wedge (H \rightarrow E1) \vee \neg G1$

4.4 Instantiating rules

So far I've dealt with fitting a formula to a formula scheme: find the principal operator; that determines the scheme which it matches; read off the way that the parameters of the scheme match the parts of the formula surrounding the principal operator.

Rule instantiation needs a bit of care. If you try to instantiate $\wedge$ intro, for example, putting $E \vee F$ for A and $G \rightarrow H$ for B, you **do not** get

$$\dfrac{\begin{matrix}\vdots & & \vdots \\ E \vee F & & G \rightarrow H\end{matrix}}{E \vee F \wedge G \rightarrow H}\ (\text{NOT} \wedge \text{intro})$$

The antecedents are all right, but the consequent is all screwed up. In the rule scheme the consequent $A \wedge B$ is a conjunction, but in the 'instance' $E \vee F \wedge G \rightarrow H$ is an implication, which brackets as $(E \vee (F \wedge G)) \rightarrow H$, and that definitely doesn't come apart into an A-part $(E \vee F)$ and a B-part $(G \rightarrow H)$.

The correct technique necessarily uses brackets to ensure that the structure of the rule's formula scheme carries across to the rule instance:

$$\dfrac{\begin{matrix}\vdots & & \vdots \\ E \vee F & & G \rightarrow H\end{matrix}}{(E \vee F) \wedge (G \rightarrow H)} \wedge \text{ intro}$$

The principle is that when inserting a formula argument into a scheme, bracket the argument if its operators would affect the structure of the scheme they are being inserted into. They affect it if they are weaker operators — a $\rightarrow$ formula inserted in an $A \wedge B$ scheme, for example — or if they are similar priority but inserted in a position which violates the binding direction — an $\wedge$ formula for B in $A \wedge B$, for example, or an $\rightarrow$ formula for A in $A \rightarrow B$.

4.5 Matching rules to formulae

Mostly we don't write down rule instances. We don't really fit formulae to rule schemes either. What we do instead is fit formulae to formula schemes by finding their principal operator. Then they fit any rule that depends on that formula scheme. It's as simple as that: the conjunction rules depend on $A \wedge B$, the implication rules depend on $A \rightarrow B$, the disjunction rules depend on $A \vee B$ and the negation rules depend on $\neg A$.

The only exception to this simple procedure is the classical contradiction rule, and I'll deal with that problem in the next chapter.

5 Proof with connectives

The target of this part of the book is neither syntax nor semantics nor the properties of Natural Deduction, it's formal proof. Proofs which use only the connectives and the constant symbols are pretty straightforward.

5.1 Stating a claim

Gentzen's invention of Natural Deduction was a spinoff from his studies of proof and proof search. One of the legacies of his work is a notation for stating logical claims. A **sequent** expresses the claim that you can prove a conclusion C from some premises $A_1, A_2, \ldots, A_n$:

$$A_1, A_2, \cdots, A_n \vdash C$$

The **turnstile** symbol $\vdash$, pronounced **proves**, is what makes the claim. If there are no premises we can miss out the turnstile, writing C instead of $\vdash C$.

Until the sequent's claim is proved it is a **conjecture**; once it's proved it is a **theorem**. Theorems can be used as auxiliary rules of inference, as we shall see.

Nowadays Natural Deduction is seen as a classification of those logics which are defined by intro and elim rules. There are lots of alternative Natural Deduction systems, distinguished by choice of rules. Even in this book there are two: the rules of Table 3.9 plus the truth rule and the constructive contradiction rule from Table 3.10 make a constructive proof system; the same connective rules plus the truth rule and the classical contradiction rule make a classical proof system.[1] The classical system can prove more theorems than the constructive. When it matters, I shall point out claims that are provable classically but not constructively.

5.2 Tree proofs

My description of rules of inference is in the style of a **proof tree**. Each proof step uses a rule to make a proof out of one or more subproofs (or, in the case of the truth rule, no subproofs at all); in doing so it makes a little **tree** of deductions,

[1] If you don't use the classical contra rule, you have a constructive proof, but just a single use of classical contra, or a theorem that depends on classical contra, makes it classical.

with the consequent of the step at the root of the tree, and its antecedents at the roots of its subtrees. That tree can be plugged as a subproof into another step to help make a larger tree, and so on. The conclusion of the whole proof is the root of the whole tree. The leaves of the tree are the premises of the proof: accept the claims made in the leaves, and the proof shows you how to persuade yourself, or someone else, to accept the conclusion written at the root.

For example, the two-premise claim

$$\begin{array}{l} \text{I am a cabbage, my name is Richard} \\ \quad \vdash (\text{I am a cabbage}) \wedge (\text{my name is Richard}) \end{array} \tag{5.1}$$

has a simple one-step proof. The proof is a valid argument — an instance of ∧ intro — but I disagree with one of the premises, so I don't have to accept the conclusion:

$$\frac{\text{I am a cabbage} \quad \text{my name is Richard}}{(\text{I am a cabbage}) \wedge (\text{my name is Richard})}\ \wedge\ \text{intro} \tag{5.2}$$

Here's an attempt to expand one of the premises above into something possibly less contentious. Now the claim has three premises:

$$\begin{array}{l} \text{I am a Martian, my name is Richard,} \\ (\text{I am a Martian}) \rightarrow (\text{I am a cabbage}) \\ \quad \vdash (\text{I am a cabbage}) \wedge (\text{my name is Richard}) \end{array} \tag{5.3}$$

Again, the proof is a valid argument, a combination of ∧ intro and → elim, but I still disagree with one of its premises!

$$\frac{\dfrac{\begin{pmatrix}\text{I am a} \\ \text{Martian}\end{pmatrix} \rightarrow \begin{pmatrix}\text{I am a} \\ \text{cabbage}\end{pmatrix} \quad \begin{array}{l}\text{I am a} \\ \text{Martian}\end{array}}{\text{I am a cabbage}}\ \rightarrow\ \text{elim} \qquad \begin{array}{l}\text{my name} \\ \text{is Richard}\end{array}}{(\text{I am a cabbage}) \wedge (\text{my name is Richard})}\ \wedge\ \text{intro} \tag{5.4}$$

This tiny tree, with only two proof steps, barely fits on the page, and careful reading is necessary to reveal its true shape. Larger tree proofs are wider still and much harder to read. So although trees are really good for explaining the rules, they aren't so good for presenting proofs. Rules that introduce assumptions make things even worse: the box mechanism isn't particularly easy to use in tree proofs, and I'm not going to try to explain it. Luckily, there is an easier way.

5.3 Line proofs

Instead of drawing a two-dimensional tree, we write a sequence of lines. Each line consists of a logical formula, numbered on the left and justified on the right.

Each line is either a premise or a deduction from a previous line (or lines). The justification either says that this line is a premise, or it names the rule used to make the deduction step and the number(s) of the previous line(s). By convention premises come first, deductions later, and necessarily the conclusion comes last. For example, a line proof of (5.1):

1. I am a cabbage	premise	
2. my name is Richard	premise	(5.5)
3. (I am a cabbage) $\wedge$ (my name is Richard)	$\wedge$ intro 1, 2	

When a proof is presented in this way it is easy to read from the top to check each step, and that's the way we usually read proofs. It's also possible to read from the bottom and trace out the equivalent proof tree, but we don't do that so often. (On the other hand, as you will see, when we make proofs we often work from the bottom upwards.)

The proof of (5.3) is now much easier to write down and read:

1. I am a Martian	premise	
2. my name is Richard	premise	
3. (I am a Martian)$\rightarrow$(I am a cabbage)	premise	(5.6)
4. I am a cabbage	$\rightarrow$ elim 3, 1	
5. (I am a cabbage) $\wedge$ (my name is Richard)	$\wedge$ intro 4, 2	

5.3.1 The line condition. In a line proof we can refer to premises or, as line 5 of (5.6) shows, previous deductions. We **must not** refer to **later** deductions, because that can produce a circular argument, one which has no proper support. Here, for example, is an invalid argument — a non-proof — purporting to show that I'm a Martian cabbage:

1. (I am a cabbage)$\rightarrow$(I am a Martian)	premise	
2. (I am a Martian)$\rightarrow$(I am a cabbage)	premise	
3. I am a cabbage	$\rightarrow$ elim 2, 4	(5.7)
4. I am a Martian	$\rightarrow$ elim 1, 3	
5. (I am a Martian) $\wedge$ (I am a cabbage)	$\wedge$ intro 4, 3	

Line 3 cheats: it deduces that I'm a cabbage from the conclusion on line 4 that I'm a Martian, which is deduced from the conclusion that I'm a cabbage, which ...and so on and on for ever. It's impossible to reorder those two lines to remove the circularity (and, indeed, it's impossible to prove the conclusion from those premises, as the methods of Part III show).

We eliminate that kind of cheating by demanding that deductions can't look downwards, in the **line condition**.

Definition 5.1 *In a line proof every line must be justified either as a premise or by use of a rule appealing to* **previous** *lines.*

5.4 Box-and-line proofs

When a logical step requires us to make an assumption, we can't draw a proof as a sequence of lines. The $\rightarrow$ intro rule, for example, lets us deduce $A \rightarrow B$ not from a proof of B, but from a proof of B given the extra supposition A: that is, from a whole subproof and not just its conclusion. Just as the proof rule uses a box, so must the presentation. The components of our proofs can be lines or boxes, the components of the boxes can be lines or boxes, the components of those boxes can be lines or boxes, and so on down, as deep as we wish to go.

In line proofs a deduction can refer to a formula on a previous line, but it can't refer to part of the formula. In box-and-line proofs, similarly, a deduction can refer to the whole of a box but not to part of it. We can't look inside boxes: they have to be taken all together or not at all. Steps inside the box, however, can look outside and appeal to earlier boxes and earlier lines.

Here, for example, is an argument justifying my belief that when I get into a bath I get wet.

1.	immersion $\rightarrow$ (liquid contact)	premise	
2.	(liquid contact) $\rightarrow$ (I get wet)	premise	
3.	(in the bath) $\rightarrow$ immersion	premise	
4.	in the bath	assumption	
5.	immersion	$\rightarrow$ elim 3,4	(5.8)
6.	liquid contact	$\rightarrow$ elim 1,5	
7.	I get wet	$\rightarrow$ elim 2,6	
8.	(in the bath) $\rightarrow$ (I get wet)	$\rightarrow$ intro 4-7	

The proof is a sequence of deductions, some simple, others not. Line 5 uses $\rightarrow$ elim: if I'm in the bath then I'm immersed (line 3) and I suppose I am in the bath (line 4). Line 6 pulls a similar trick with lines 1 and 5 to show that I'm in contact with liquid, and line 7 uses the definition of wetting (line 2) and line 6. Then the entire box from lines 4 to 7 is cited in the justification for the deduction on line 8.

5.4.1 The box-and-line condition. To make our proofs secure we have to avoid circularities, and we have to make sure that assumptions and lines deduced from assumptions aren't used outside their scope. This is the **box-and-line condition**.

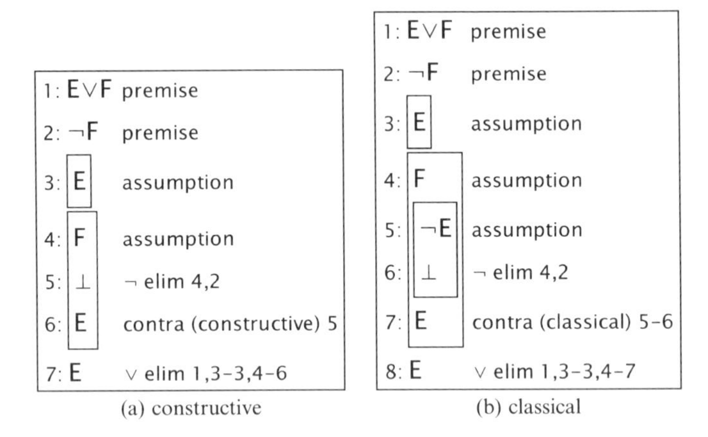

(a) constructive (b) classical

Fig. 5.1 Constructive and classical proofs of $E \vee F, \neg F \vdash E$

Definition 5.2 *In a box-and-line proof*

1. *every line must be justified either as a premise or by use of a rule appealing to* **previous** *lines or boxes;*
2. *if an appealed-to line is inside a box, then that box must also enclose the justified line.*

The condition can be summarized in the slogans of Table 5.1.

Table 5.1 The box-and-line slogans

• no looking downwards;
• no peeking inside a box.

5.4.2 Justifying lines in box-and-line proofs. There are two kinds of antecedent in the rules I've given in this book. **Unboxed antecedents**, like those in the $\wedge$ intro rule (see (5.5) and (5.6)), are always an earlier **line**, and I refer to that line with its number. **Boxed antecedents**, which introduce an assumption, like the one in $\rightarrow$ intro (see (5.8)), are always an earlier **box**, and I refer to that box with two numbers i-j, meaning the box which starts on line i and ends on line j.

It's that simple. A step which appeals to $\vee$ elim, the most complicated connective rule, for example, always refers to three antecedents: a line $(A \vee B)$ and two boxes: one with assumption A and last line C; the other with assumption B and last line C. Fig. 5.1(a), for example, is a formal proof made in Jape of the claim $E \vee F, \neg F \vdash E$ which was loosely discussed in Section 3.6.1. The only oddity is the single-line box on line 3: that's what a proof of a conclusion from an identical assumption looks like! The important step is the one on line 5 which appeals to lines 4 (assumption F) and 2 (premise $\neg F$) to derive a contradiction, and hence allows us to dismiss that arm of the $\vee$ elim proof.

Fig. 5.1(b) is the classical version of the same argument, to illustrate how it's always possible to imitate a constructive contradiction step at the cost of introducing an unnecessary assumption. Note that the extra assumption, on line 5, is not appealed to anywhere.

> Justifications of deduced lines must respect the antecedent-ordering of their corresponding rules. On line 5 of Fig. 5.1(a), for example, "$\neg$ elim 4, 2" means that line 4 is A, line 2 is $\neg A$, because that is the order those antecedents appear in the $\neg$ elim rule (see page 43). "$\neg$ elim 2, 4" wouldn't make sense because line 4 isn't $\neg\neg F$, as it would have to be if it is $\neg A$ and line 2 is A.

5.5 Real-life proofs with formal rules

Formal logic started as an attempt to distinguish valid from invalid reasoning in real world disputes. Logic is relevant to real life, but it's hard to apply it, and this book isn't about the fit between logic and everyday reasoning. Nevertheless it's worthwhile looking at some real-world examples to get a feel for the way that logical rules work.

Some of the reasoning which follows is good, and some is bad. I consider bad reasoning because one way to get a feel for the rules is to see situations in which they **don't** work.

5.5.1 The warm room. I like to be warm. I often squabble with Bernard Sufrin, the colleague and friend with whom I first developed Jape, because he likes things to be cooler. I think he's a polar bear; he thinks I'm a softie. The argument has been going on for decades. This is the way **I** see it:

> "You say it's too hot. You know I always fall asleep when it's too hot. But I'm awake now, so it can't be too hot. Please leave the thermostat alone!"

I prove to my own satisfaction, using a valid argument, that it isn't too hot. My argument has premises and uses $\neg$ intro, $\rightarrow$ elim and $\neg$ elim. Relying on the conclusion of that argument, I shamelessly try to impose my will on Bernard, in his own room.

When I claim "it can't be too hot", I really mean that Bernard can't reasonably claim that it is too hot: I think I've proved it's impossible for anybody to agree that it's too hot — i.e. $\neg$(too hot). Looking at the rules summarized in Table 3.9 on page 43, the ¬ intro rule seems to show a way to prove such a conclusion: ask the opponent (in this case, Bernard) to suppose that it is too hot, and try to show a contradiction.

$$\dfrac{\boxed{\begin{array}{c}\text{too hot}\\ \vdots \\ \bot\end{array}}}{\neg(\text{too hot})}\ \neg\ \text{intro} \tag{5.9}$$

Formal logical arguments start with a statement of premises. Real-world arguments introduce them as and when they are needed. My dispute with Bernard rests partly on a claim about my weakness: I do tend to fall asleep when I feel warm. In the argument that premise is the sentence "You know I always fall asleep when it's too hot". I might capture this assertion with the logical claim (too hot) → (I am asleep). I use this premise together with the assumption in a step of → elim:

$$\dfrac{(\text{too hot}) \to (\text{I am asleep}) \quad \text{too hot}}{\text{I am asleep}}\ \to \text{elim} \tag{5.10}$$

But then — contradiction! — my other premise is that I'm awake, which I take to mean ¬(I am asleep):

$$\dfrac{\begin{array}{c}\vdots\\ \text{I am asleep}\end{array} \quad \begin{array}{c}\vdots\\ \neg(\text{I am asleep})\end{array}}{\bot}\ \neg\ \text{elim} \tag{5.11}$$

I can put all the steps together in a box-and-line proof:

1.	(too hot) → (I am asleep)	premise	
2.	I am awake	premise	
3.	too hot	assumption	
4.	I am asleep	→ elim 1, 3	(5.12)
5.	⊥	¬ elim 4, 2	
6.	¬(too hot)	¬ intro 3-5	

This 'proof' won't stop us squabbling. It isn't true that I always fall asleep when I'm too warm, so Bernard can attack the premise on line 1. Or perhaps I am not as awake as line 2 claims. Most effectively, he can point out that I'm equating "too hot" on line 1 with "too warm for Richard". All I've really proved is that the room isn't too warm for me, so our tussles round the thermostat will continue!

5.5.2 The dead Major. In life we often reason badly because to survive in a dangerous world we must be prepared to extrapolate from scraps of evidence. When we hear a rustle in the undergrowth we have to jump to conclusions before the tiger jumps on us: jumping too soon may be a logical error, but jumping too late can be fatal. Even our most confident extrapolations don't amount to logical proof: we don't **know**, for example, that the sun will rise tomorrow — that is, we have no proof — but we plan our lives as if we do know. Given the long recorded history of repeated sunrise, we'd be foolish not to. But gambling is always risky: as Damon Runyon put it, if somebody bets you that he can make the Jack of diamonds jump up and squirt cider in your ear, don't take the bet because for sure if you do, you will get an ear full of cider. (I'd still bet on sunrise, though.)

Lots of detective thrillers tempt us to jump to a conclusion, perhaps like this:

> *"Submersion in water causes drowning; drowning causes death; the Major was found dead in the lake; the Major was drowned!"*

Translated into logical notation, the argument goes as follows:

1. submersion → drowning	premise	
2. drowning → death	premise	
3. submersion → death	from 1, 2	
4. submerged	fact	(5.13)
5. dead	fact	
6. drowned	from 4, 5, 1	

The Major was drawn dripping wet and obviously dead from the lake. It seems he must have been drowned, and that looks like a logical conclusion from the evidence. Unfortunately for our pride in our deductive skills, we are told in the last chapter that the Major was poisoned by the butler and then thrown in the lake. He wasn't drowned at all!

If the conclusion of a logical argument is untrue, there must be something wrong with the premises, or the reasoning, or both. Perhaps our 'proof' is mistaken because it doesn't follow the rules: lines 3 and 6 look like deduction steps, but aren't justified by the rules of Chapter 3; lines 4 and 5 aren't labelled as premises or the result of deduction steps.

Gaps and slips in reasoning often happen when people are trying to reason **semantically**, using meanings instead of formula shapes. We know, or we think we know, a good deal about the likely causes of death in imaginary 1930s country houses, and we correspondingly rush to judgement. Semantic reasoning is useful and sometimes essential in real life (tigers and all that) but formal (shapewise) reasoning is more reliable when we can use it. Let's see if we can make a formal proof out of what we've been given.

First, lines 4 and 5, labelled as 'fact', are really two extra premises for the argument. It's best to list the claims on which a proof depends as explicit premises: then we can see what we are relying on, and decide whether we want to rely on them or not. The proof, if it is a proof, should start:

1. submersion $\rightarrow$ drowning	premise	
2. drowning $\rightarrow$ death	premise	
3. submerged	premise	(5.14)
4. dead	premise	
5. ...		

Line 3 of (5.13) claims that submersion $\rightarrow$ death follows from the premises submersion $\rightarrow$ drowning and drowning $\rightarrow$ death. This not a single-step deduction according to the rules, but it is logically valid.

> If I accept $E \rightarrow F$ and $F \rightarrow G$ then when I accept E I must accept F; then, because I accept F and $F \rightarrow G$ I must accept G; so if I accept E I accept G; that is, I accept $E \rightarrow G$.

Formal reasoning follows the same track, and here's a proof made in Jape:

1:	E→F	premise	
2:	F→G	premise	
3:	E	assumption	
4:	F	→ elim 1,3	(5.15)
5:	G	→ elim 2,4	
6:	E→G	→ intro 3-5	

This proof establishes the theorem $E \rightarrow F, F \rightarrow G \vdash E \rightarrow G$. Because it uses generic formula-names E, F and G rather than specific formulae like drowning, submersion and death (ugh!) we can use it, instantiated with particular formulae — submersion for E, drowning for F, death for G — just as if it was an extra logical rule:

1. submersion $\rightarrow$ drowning	premise	
2. drowning $\rightarrow$ death	premise	
3. submerged	premise	
4. dead	premise	(5.16)
5. submersion $\rightarrow$ death	$E \rightarrow F, F \rightarrow G \vdash E \rightarrow G$ 1, 2	
6. ...		

The last step in (5.13) was to conclude that the Major was drowned. That's logically ok, as it turns out: submersion causes drowning; the Major was clearly submerged (look at the corpse! still dripping wet!) so surely he was drowned:

1. submersion → drowning	premise	
2. submerged	premise	(5.17)
3. drowned	→ elim 1, 2	

It turns out that the original argument was far too complicated. Three lines are all that we need: two premises, one straightforward deduction, a valid and convincing argument. Although the original argument was padded out with too many premises and some unnecessary deductions, the conclusion seems to be a logical necessity. Then we find out that the Major **wasn't** drowned: the butler poisoned him. What went wrong?

If the conclusion of a valid argument contradicts reality — as this one seems to, provided you believe the detective's claims about poisoning — there must be something wrong with the premises. The Major was clearly submerged, so we can't attack that premise. But if you submerge a corpse, it doesn't drown — you just get a wet corpse! Only live things drown: submersion doesn't **necessarily** cause drowning. The argument should have been stated like this:

1. submersion ∧ (alive when submerged) → drowning	premise	
2. submerged	premise	
3. alive when submerged	premise	(5.18)
4. submersion ∧ (alive when submerged)	∧ intro 2,3	
5. drowned	→ elim 1, 4	

Still we reach an untrue conclusion. But this time we can't be confounded when it's revealed that the butler **first** killed the Major and **then** threw him into the lake: the premise on line 3 isn't valid, so the reasoning is irrelevant and we can forget about the conclusion. Instead we can be cross with the author for misleading us with all that stuff about the detective's efforts to revive a drowned man!

5.5.3 Cars and congestion. The UK has too much traffic for its roads, and most of the traffic is cars. I and several million others live in London, where the traffic problem is specially acute despite the fact that it has the best public transport provision in the country. This sparks lots of hot political argument, with alternative causes constantly being suggested and novel remedies put forward. You'll be relieved to know that people try to be logical when arguing about such a controversial issue.

Sometimes our wish to believe in a particular causal relationship makes us accept a bad argument if it's on our side. As a cyclist, I'm a sucker for arguments that 'prove' cars are a Bad Thing, and I fell for the one that follows.[2]

Some people — call them Greens — think that it is bad for UK children to be driven to school. Casting about for leverage which might encourage parents to let their children walk or cycle or take a bus,[3] they argue that '**school run**' traffic adds to congestion in the morning rush-hour. Their argument can be summarized as follows:

> *"On normal weekday mornings there is congestion. During the school holidays, and at half-term breaks, there is much less congestion. Clearly school-run journeys are causing congestion."*

Most of this summary is evidence: appeal to experience of the outside world, premises that we can verify by experiment. The logical reasoning is entirely hidden, and it's my job to reconstruct it so that we can check it.

Evidence about the relationship between school runs and congestion is non-controversial. All the traffic authorities and all the disputants agree that it happens more on school days, just as the Greens claim, so we can't attack their argument at that point. If there's anything wrong it must be to do with other assumed and unstated premises and/or the logical reasoning which leads to their conclusion.

To lay bare the logical reasoning, I begin by stating the evidence-claims as logical implications:

$$(\text{school day}) \rightarrow \text{congestion} \qquad \neg(\text{school day}) \rightarrow \neg\text{congestion}$$

These formulae don't capture the original claims precisely — Greens don't claim that there's **no** congestion on non-school days — but my version fits the Natural Deduction rules more easily than the claims they make. If we pretend to accept the absolute congestion claims we shall reveal the bones of a logical argument, and we won't miss anything important.

We haven't finished with premises yet, because the argument doesn't explain how school-run journeys are connected to school days. The Greens assume that everybody listening to them would realize that school-run journeys take place only on school days. That tacit knowledge can be revealed in two more premises:

$$(\text{school day}) \rightarrow (\text{school run}) \qquad \neg(\text{school day}) \rightarrow \neg(\text{school run})$$

You might quibble that this too is an over-simplification, because there will always be a few people who forgetfully make a school run on a non-school day.

[2] Bad arguments don't disprove logical claims. I still think London has too many cars.

[3] We don't have a school bus system in London, so pupils have to ride on the public buses with everybody else. Some parents don't like that.

It's hard to dance the complexities of reality in the lead boots of logic, but I don't think I'm over-simplifying if I equate 'school-run journeys' with 'journeys that actually transport children to school'. There are lots of journeys of the second kind on school days, surely none at all on non-school days. A tiny minority of absent-minded people will do daft things every day, and we can safely ignore them.

Given those four premises, the Greens might expand their argument as follows.

> *"Suppose there's a school run; then it must be a school day; and there is congestion on school days; so when there's a school run there is congestion.*
> *"Suppose that there is congestion; congestion only happens on school days, and on school days there will be a school run; so when there is congestion there is a school run.*
> *"Similarly, when there isn't a school run there isn't congestion, and vice versa.*
> *"Therefore, school runs cause congestion."*

If it is a proof, it looks like it is trying to be a proof by logical argument. So: is it a valid argument? Well, yes it is, but it's not yet an argument by the rules, because there are gaps which have to be filled in. Take the very first step: "suppose there's a school run; then it must be a school day". What's the logical step? It isn't a step of $\rightarrow$ elim from the premises: we have (school day) $\rightarrow$ (school run) and $\neg$(school day) $\rightarrow$ $\neg$(school run), both constructed on the sensible basis that school days cause school runs; we **don't** have a premise (school run) $\rightarrow$ (school day).

If we don't have that premise then perhaps we can prove that it is a consequence of the premises we do have. Perhaps surprisingly, it is! Indeed in general, when I accept $\neg F \rightarrow \neg E$ and I also accept $F \vee \neg F$, I must accept $E \rightarrow F$. The informal argument goes as follows:

> I accept $\neg F \rightarrow \neg E$. Suppose for the sake of argument that I accept E: I can't also accept $\neg F$ because then by implication I would have to accept $\neg E$, and that would be a contradiction with my assumption. But I do accept $F \vee \neg F$, so if I can't accept $\neg F$ I must accept F. Summarizing, if I accept E I am forced to accept F: that is, I accept $E \rightarrow F$.

Formal reasoning mimics this argument. Classical reasoners tacitly accept $F \vee \neg F$ and build it into their treatment of contradiction. A constructive proof would demand an extra premise to be sure that either it's a school day or it isn't, which in this case seems a bit unnecessary so, for the sake of simplicity, I'll ignore constructivist squirmings. Here's the formal classical proof, done in Jape:

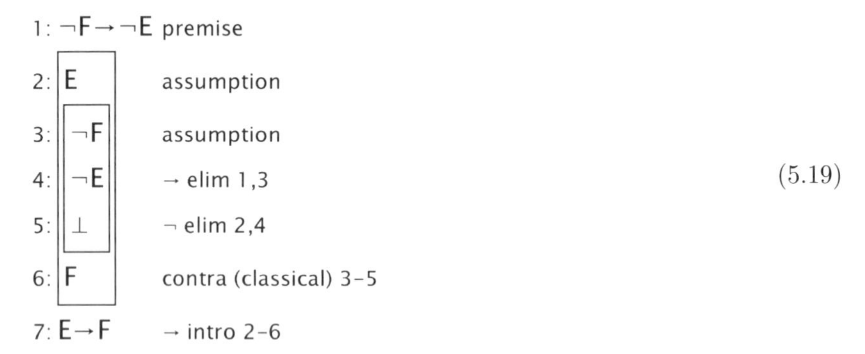

(5.19)

The box from line 2 to line 6 shows that if you accept E you must accept F. It does it by using a subsidiary box to show that if you accept $\neg F$ there's a contradiction. Classical reasoning concludes that if you can't accept $\neg F$ you must accept F, and that's the end of it. Overall, the proof establishes a logical theorem: $\neg F \to \neg E \vdash E \to F$, and, instantiating that theorem, we can deduce (school run) → (school day) from the premise ¬(school day) → ¬(school run).

> At this point I draw your attention to something going wrong with the attempt to prove a cause by logical reasoning. School runs do **not** cause school days! A proof that (school run) → (school day) merely shows persistent association. Association isn't cause. Nevertheless, I press on.

With the theorem of (5.19) to aid me, I can make a complete proof that from agreed facts about the world it follows that (school run) → congestion:

1.	(school day) → congestion	premise	
2.	¬(school day) → ¬(school run)	premise	
3.	school run	assumption	
4.	(school run) → (school day)	$\neg F \to \neg E \vdash E \to F$ 2	(5.20)
5.	school day	→ elim 4,3	
6.	congestion	→ elim 1,5	
7.	(school run) → congestion	→ intro 3-6	

With a little more difficulty, because it means playing around with negation, you can show that ¬(school run) → ¬(congestion) (I'll leave the details of the proof to you) and you can also show the other two logical claims made in the Greens' argument.

Given those proofs we have to face the Greens' final claim. When the 'cause' happens the effect happens; when the 'cause' stops the effect stops; when the effect happens the 'cause' is operating and when the effect stops the 'cause' has stopped. Is that enough to prove that the effect is **because of** the 'cause'? Aren't we now sure that the school run should be outlawed?

Certainly not.

5.5.4 An alternative cause. Proving an implication $A \rightarrow B$ merely proves a particular sort of association between A-events and B-events: when A happens, so does B. Association doesn't prove cause — hence lots of long-running disputes about smoking and cancer, soft and hard drug use, vaccination and disease, social depravity and the collapse of empires, and so on and on.

Logical implication isn't about causes, but scientific implication is. To prove a scientific implication you have to demonstrate an association of cause and effect **and** describe a convincing **mechanism** which connects the two. But scientific proof is contestable if others can show that some of the parts of the mechanism can't be found, or can propose a simpler mechanism, in which case further research is needed to find out which mechanism is actually operating.

One attack on the Greens' argument, then, is to suggest that some cause that they have overlooked might be causing the observed congestion effects. Some opponents of the green position press just that point:

> *"Certainly the school run days are the congestion days, and vice versa. But the effect is caused by school holidays. Parents go on holiday when their children are on holiday. There are fewer drivers travelling to work, and there is therefore less congestion, during the school holidays."*

This is a perfect bit of scientific counter-argument: "your proposed mechanism can't be considered a cause until you have disproved this plausible, and apparently simpler, alternative". Until we can be sure that the alternative mechanism — school holidays means fewer drivers means less congestion — isn't the explanation, the green argument won't get the acceptance that perhaps it deserves.

I don't know if traffic planners take the school-holiday argument seriously, but it doesn't seem as if anybody has yet done the expensive surveys of people in traffic jams, people at work and people on holiday to find out which of the proposed causes is operating. So nobody really knows whether the school-holiday argument is a better explanation for congestion than the school-run argument.

5.5.5 Small cause, small effect? Another attack tries to proceed by a form of contradiction, arguing that the effect can't be caused in the way that the Greens suggest. (This is a real example, straight off the radio, honestly it is!)

> *"My organization has done a survey. We have found that the school run can't be having a large effect on traffic, because less than 10% of children are driven to school."*

The principle which seems to be being used in this argument is that small causes must have small effects. That's debatable, of course, and in the limit it's paradoxical, when the last straw breaks the camel's back. In road traffic we find back-breaking straws everywhere,[4] but I set that objection aside for the time being. **If** a small percentage increase in road traffic has only a small effect on congestion, and **if** school-run journeys are a small proportion of all morning rush-hour journeys, **then** we would be forced to agree that school-run journeys can't be having a large effect on congestion.

But that isn't the claim that is being made. The claim is that school-run journeys are a small proportion of **travel-to-school** journeys. What is the relationship of school-run journeys to **morning rush-hour car** journeys? There are hidden premises here. The presenter of this argument may have thought that it was obvious that, since children are a minority of the population (agreed) and since most journeys to work are by car (not agreed — certainly untrue in London where public transport takes almost all the strain), travel-to-school journeys must be a small proportion of rush-hour car-to-work journeys. It seems possible that in London school-run journeys are a significant proportion of rush-hour car journeys, and if so the counter-argument would be demolished.

5.5.6 Whose fault is it, then? Despite all the attempts at logical proof, we still don't know whether or not the school run has a major or a minor effect on morning rush-hour traffic congestion. Logical reasoning, however careful, isn't enough to expose causes but it can expose the points at which research is worthwhile and where there can be rational dispute.

5.6 Searching for formal proofs

This book is about formal — shapewise, meaningless — proof, and not about finding formal correspondences with real-world reasoning. It is about **making** proofs, as well as reading them. It's time to turn our back on semantic reasoning and the difficulties of sleepy academics, dead Majors and congested streets and turn to the making of formal proofs.

Formal proofs are quite easy to read, once you understand the logical rules, but they do look very inventive: when you first see one you wonder how the steps were chosen and how the assumptions were dreamed up. In practice, because proofs aren't found in the way that they are read, making proofs is mostly a kind of rational search and only rarely a species of invention. The proof calculator

[4] The whole of central and east London was once paralysed for several hours because a single lorry broke down in a river tunnel. The police had to smuggle the lorry driver away and keep his name and address secret for fear of reprisals. I got home smug and warm and as quickly as usual on my bike, of course.

Jape, available on the internet for free (see www.jape.org.uk), is designed to help you to learn about proof search by playing with real, if small, formal proofs. Once you've learnt proof search strategy by playing around in Jape, you can use it on paper and on blackboards and in examinations. And the strategy is really, really, **really** easy to use.

That strategy is summarized in Table 5.2. Proof search is driven by the shapes of formulae. It's pretty straightforward, provided you can overcome what seems to be an inborn novice prejudice against searching backwards from a conclusion. There are further slogans about particular connectives and their rules, but these are the core of what you need to know. Only slogan 8 looks weird (classical contradiction gets special treatment in Section 5.6.6).

Table 5.2 Slogans for Proof Search

1. Don't always work forwards.
2. Work on formulae with connectives.
3. Shape-match with formula-schemes in rules.
4. Fit hypothesis formulae to antecedents of elim rules; fit conclusion formulae to consequents of intro rules.
5. Elim steps usually go forward, intro steps nearly always go backward.
6. Prefer rules that generate assumptions.
7. Believe in slogan 1.
8. If all else fails, classical contradiction might be worth a try.

5.6.1 Searching for proofs in Jape. Jape is a proof calculator. That means it makes proof steps, just as an arithmetic calculator makes arithmetic steps. As a calculator, it doesn't give you any help: the step you choose to make may not be one that leads to a proof. But, unlike an arithmetic calculator, it can undo a step or several steps, so you can use it to search through the maze of possible proof developments. And, as a calculator, it guarantees accuracy, so you can use it to find out just what the effect of a proof step would be. The manual that comes with Jape tells you how to drive it with the mouse and the keyboard, and I shan't repeat that information here.

What makes Jape useful is that it can help you to understand and learn a strategy for **finding** proofs. After Chapter 4 you understand how rules fit formulae; once you've learnt a few of the basic moves from the Jape manual, you can begin to play.

The proofs you've seen so far in this book, apart from some bits of proof about the undrowned Major, have been complete. Jape deals with complete proofs and also with **proof attempts**: box-and-line structures in which all the

1: E∧(F∧G) premise
...
2: (E∧F)∧G

1: E→F, F→G premises
...
2: E→G

...
1: (E→F)∨(F→E)

Fig. 5.2 Three sample conjectures as seen in Jape

deductions that have been made are valid logical steps, but which need more deductions to complete them.

You begin a proof attempt by choosing a conjecture from one of Jape's panels, or by typing in one of your own. What you see, as for example in Fig. 5.2, is immediately a proof attempt, with a line of dots separating the conjecture's premises from its conclusion and showing the logical gap which has to be bridged; even if there are no premises the dots are still there, because there's still logical work to be done. To save precious screen space, Jape usually puts as many premises as it can on a single line. You work forwards from formulae above the dots — called **hypotheses** — and/or backwards from a formula immediately below a line of dots — called an **open conclusion**.

5.6.2 Proof search with $\wedge$. Consider the formula $E \wedge F \wedge G$. The priority rules of Table 4.3 tell us to read it as $((E \wedge F) \wedge G)$, but the meaning of $\wedge$ suggests that it doesn't matter how we bracket or order conjunctions. We might hope to be able to make a formal proof which supports that intuition. Actually we would need two proofs: one of $E \wedge (F \wedge G) \vdash (E \wedge F) \wedge G$ and another, the other way round, of $(E \wedge F) \wedge G \vdash E \wedge (F \wedge G)$. If you accept one formula you can prove the other; if you accept the other you can prove the one; so they are equivalent.

Fig. 5.3 shows the stages of a formal proof of half of the equivalence, found by following the strategy slogans of Table 5.2. Slogan 2 doesn't direct the first step, because both the premise on line 1 and the conclusion on line 2 include connectives. Slogan 3 tells us to shape-match; slogan 4 tells us what formula-schemes to match. Line 1 fits either of the $\wedge$ elim rules, and line 2 matches the $\wedge$ intro rule: we therefore have a choice of backward or forward step, but since none of the matching rules generates an assumption, slogan 6 doesn't apply. Perhaps we should prefer the backward step, because slogan 1 suggests that forward steps aren't always the right thing.

Despite slogan 1, most people would make a forward step in this situation just because they can, and Fig. 5.3(b) shows the effect of $\wedge$ elim applied to line 1 of Fig. 5.3(a). I chose the rule which deduces A from $A \wedge B$: it's then reasonable to apply the other elim rule to the same premise (Fig. 5.3(c)). Line 2 of Fig. 5.3(c) is a new hypothesis which includes a connective, and Fig. 5.3(d)

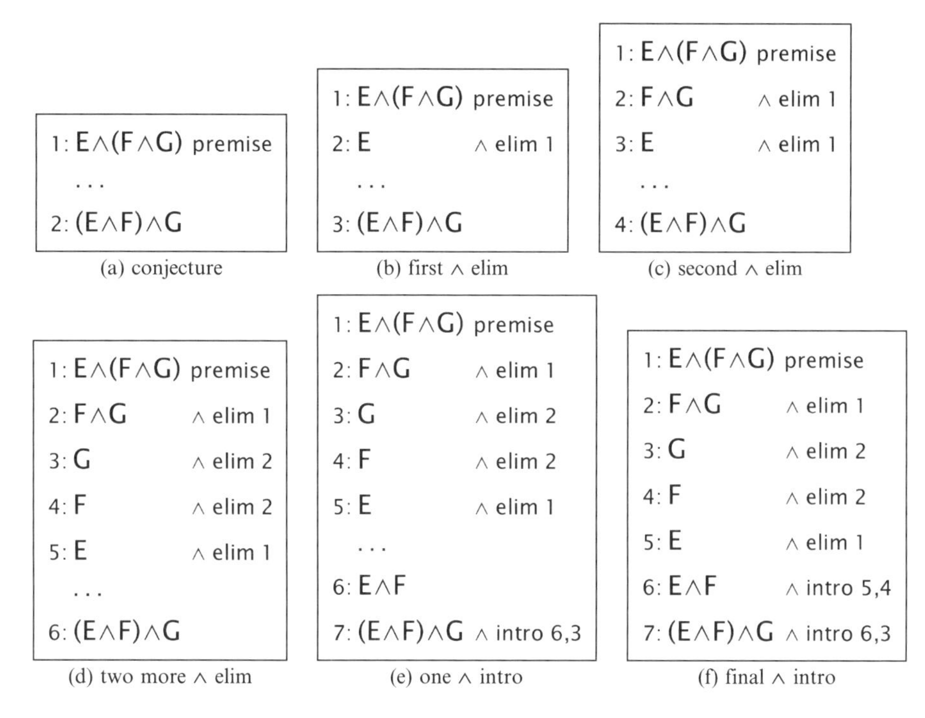

(a) conjecture (b) first ∧ elim (c) second ∧ elim

(d) two more ∧ elim (e) one ∧ intro (f) final ∧ intro

Fig. 5.3 A proof using ∧ rules

shows the result of applying two more elim steps to that line. The effect so far is to have extracted all the component parts of the premise.

Now that the connectives in the hypotheses are used up — all we could do is produce more copies of the lines we've already made — slogan 2 tells us that there's nothing for it but to match line 6 of Fig. 5.3(d), the only active conclusion. An $\wedge$ intro step generates two antecedent conclusions in Fig. 5.3(e): one ($E \wedge F$) is shown as the new line 6, but the other (G) already appears on line 3, so Jape appeals to it, as it should.

Now line 7 is no longer open, and all there is to work on is the new line 6. Another $\wedge$ intro step links it immediately to lines 5 and 4, and the proof is complete in Fig. 5.3(f).

It's all very straightforward if you follow the slogans. If it weren't for slogans 1 and 5, you might be tempted in Fig. 5.3(e) to try to make $E \wedge F$ from lines 5 and 4, making an intro step forward. That's possible in Jape but it's much harder than working backwards from line 6. That's what slogans 1 and 7 are about!

5.6.3 Proof search with →. Implication usually involves backward proof search, much to the consternation of novices. But sometimes it's possible to get away

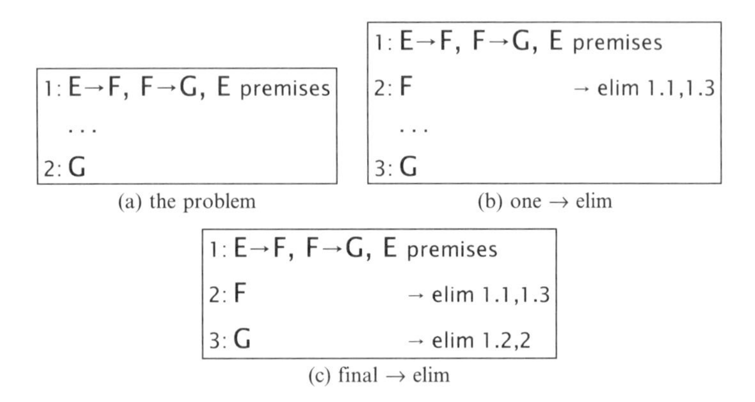

(a) the problem (b) one → elim

(c) final → elim

Fig. 5.4 Forward proof search with →

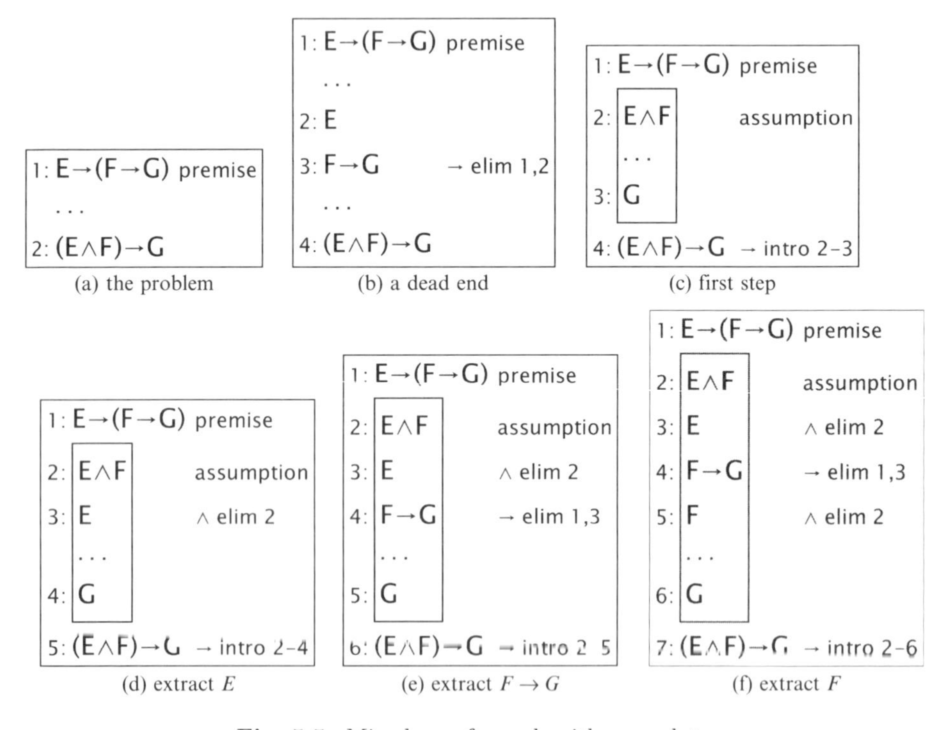

(a) the problem (b) a dead end (c) first step

(d) extract E (e) extract $F \to G$ (f) extract F

Fig. 5.5 Mixed proof search with → and ∧

with forward steps only, as in Fig. 5.4. Slogan 4 is the only one we need in this case.

More often we need a mixture of strategies, as in Fig. 5.5. The timid prover, tempted to try → elim as a first step, only drives up a dead end (Fig. 5.5(b)).[5] The correct first step, guided by slogans 1 and 6, is → intro, which produces

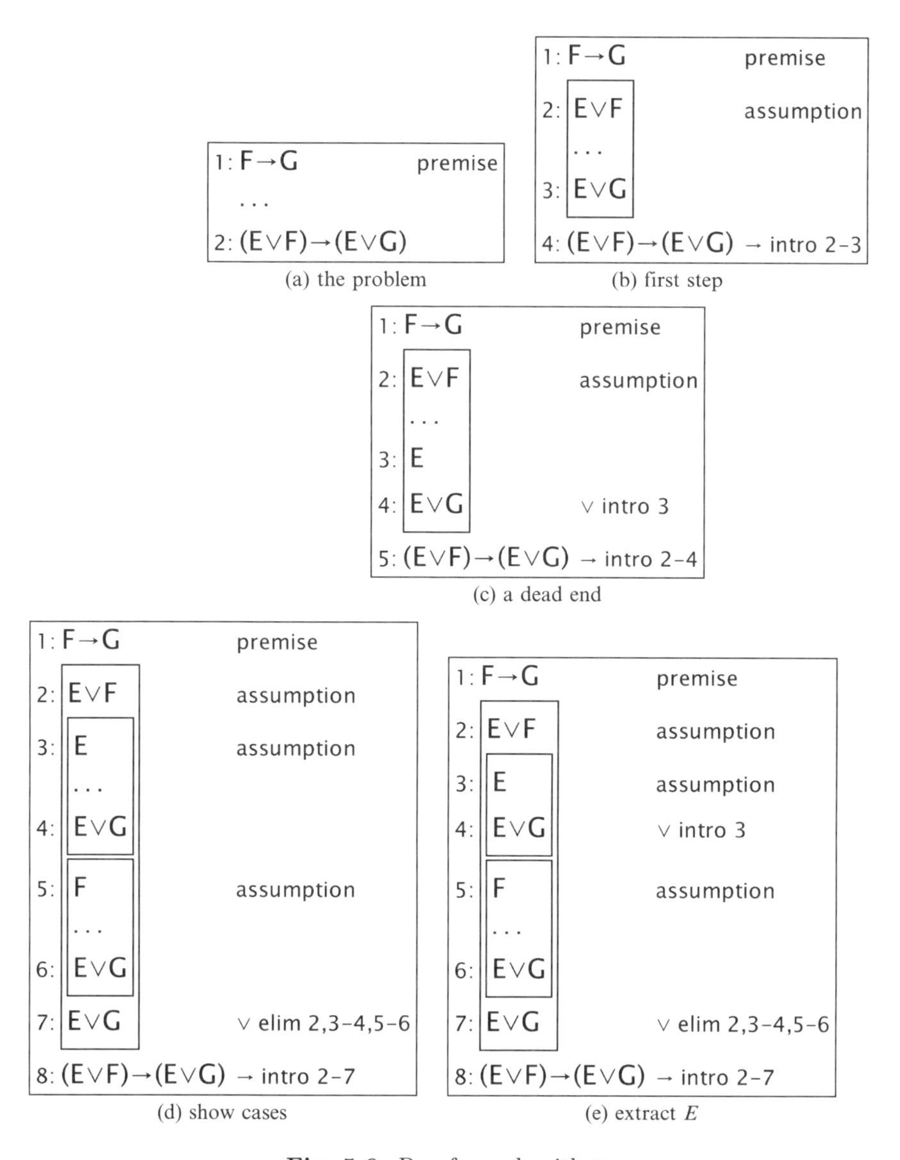

(a) the problem

(b) first step

(c) a dead end

(d) show cases

(e) extract E

Fig. 5.6 Proof search with ∨

the box in Fig. 5.5(c). The rest is forward reasoning, first extracting E, then $F \to G$, then F, and finally G (not shown).

[5] Why is it a dead end? Simply, because you can't prove it's raining when all you know is that **if** it's raining, **then** you will get wet: knowing $A \to B$ doesn't tell us A. More on this in Chapter 8.

5.6.4 Proof search with $\vee$. For every novice who finds backward search with $\rightarrow$ intro disturbing, there must be five who think, the first time they try it, that forward search with $\vee$ elim is extremely dangerous. Certainly, it makes a big change in the display, and splits the proof search into two: but that's what argument by cases **is**.

Presented with the problem in Fig. 5.6(a), and fortified by previous experience with implication, most people can pluck up enough courage to attempt a backward $\rightarrow$ intro step, giving 5.6(b). Then the step which produces the least change in the picture, but which leads to a dead end, is $\vee$ intro backwards. There are two versions of the step: Fig. 5.6(c) shows the one which extracts E from the conclusion; the one which extracts G is just as bad.

The problem is that $\vee$ intro, in focussing on half the conclusion, throws the other half away, and that's usually not a good idea early on in a proof. The rule has a slogan all of its own.

Table 5.3 $\vee$ intro slogan

$\vee$ intro resolves uncertainty; use it **as late as possible.**

The proof has to involve both sides of the conclusion $E \vee G$: neither of them is irrelevant; neither can be thrown away yet. The correct second step, directed by slogan 6, is $\vee$ elim forwards, which gives Fig. 5.6(d). The search has been split into two cases, shown by the two boxes each with its own lines of dots. Now it's possible to see how to use $\vee$ intro properly: in one case we can easily prove E, in the other G, and we can then resolve the $E \vee G$ uncertainty differently in the two different cases. The proof in one case is immediate (Fig. 5.6(e)); the search in the other case is straightforward but not shown.

5.6.5 Proof search with $\neg$. Fig. 5.7 shows a proof of $E \vee F \vdash \neg(\neg E \wedge \neg F)$, half of one of de Morgan's Laws, an equivalence familiar to anybody who has studied Boolean arithmetic or computer hardware design. In Fig. 5.7(a), slogan 6 doesn't tell us how to choose between backward $\neg$ intro and forward $\vee$ elim, since each introduces an assumption (or assumptions). Either approach will work, but I've chosen to do the $\neg$ intro first, since exploration shows that it produces a slightly shorter proof. Once both those steps are complete (Fig. 5.7(c)) you can see that E on line 3 contradicts $\neg E$ in line 2: extract $\neg E$ (5.7(d)), make the contradiction (5.7(e)) and the first case is closed. The second case is similar, and not shown.

Although this proof uses the contradiction **symbol**, it doesn't use a contradiction **rule**. It's therefore definitely inside the blob of Fig. 3.3, a proof which both constructivists and classicists can happily accept. It's a straightforward connective-driven proof.

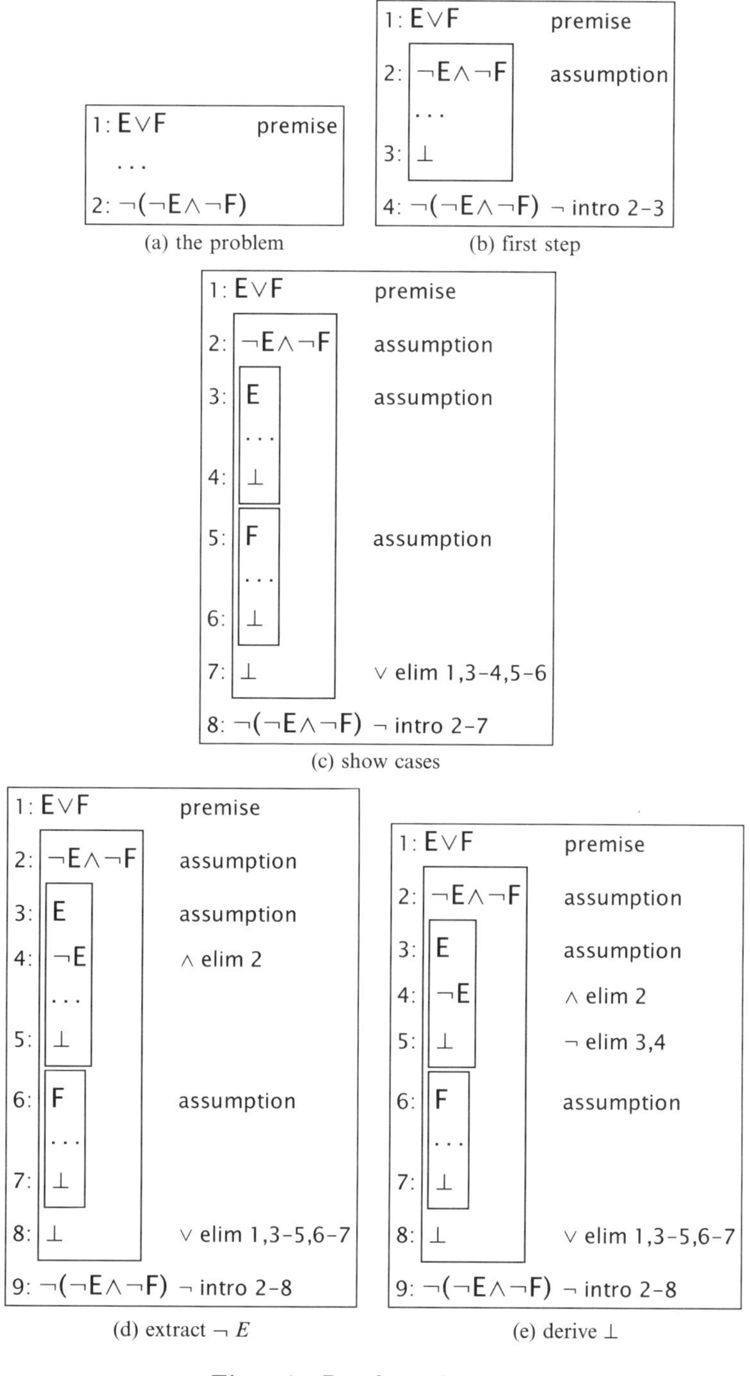

(a) the problem

(b) first step

(c) show cases

(d) extract $\neg E$

(e) derive $\bot$

Fig. 5.7 Proof search with $\neg$

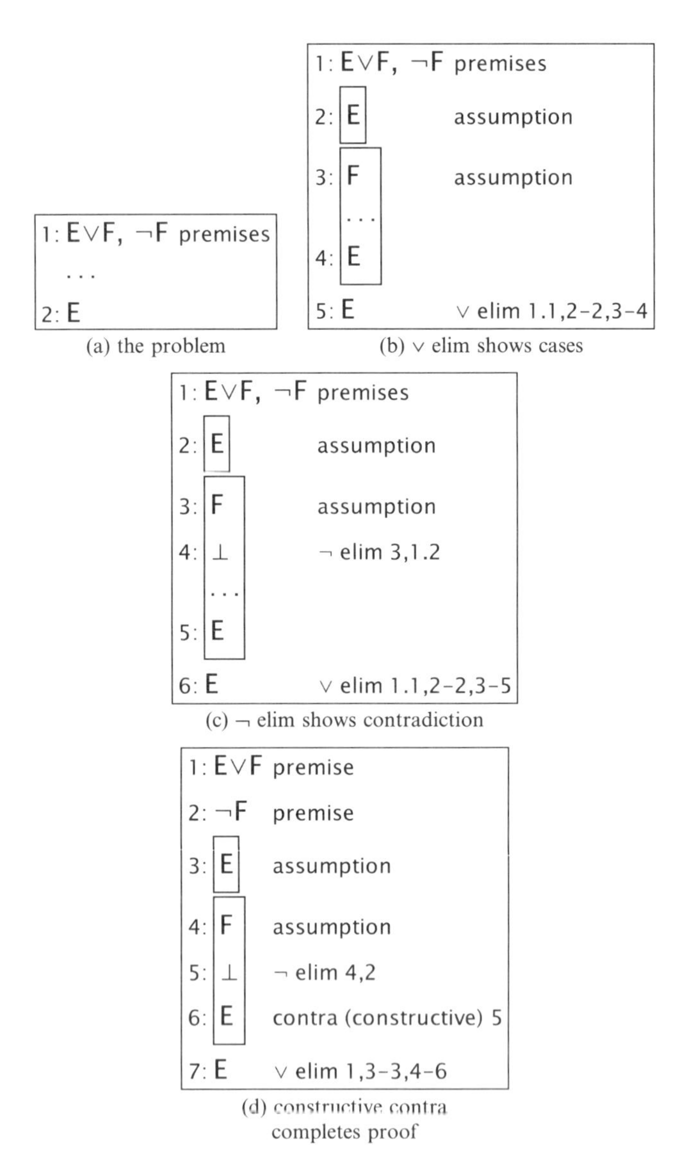

Fig. 5.8 Proof search with constructive contradiction

5.6.6 Proof search with $\bot$. Contradictions introduced by $\neg$ intro form the target of $\neg$ elim. You usually use $\neg$ elim to produce a contradiction symbol and then constructive contradiction gives your conclusion, as illustrated in Fig. 5.8, but you can do it the other way round if you want to. Constructive contradiction is **really** easy to use.

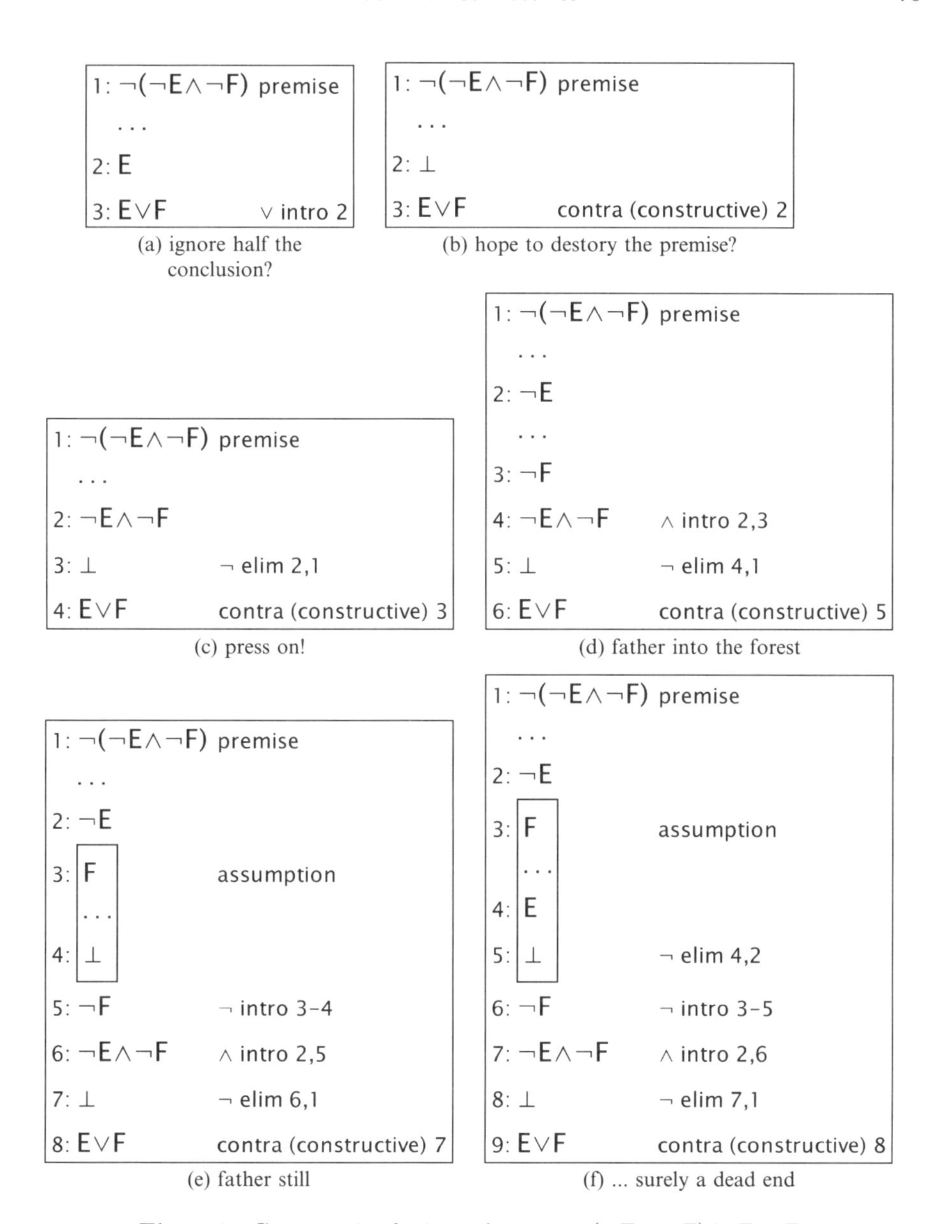

(a) ignore half the conclusion?

(b) hope to destory the premise?

(c) press on!

(d) father into the forest

(e) father still

(f) ... surely a dead end

Fig. 5.9 Constructive logic can't prove $\neg(\neg E \wedge \neg F) \vdash E \vee F$

But easy steps aren't always enough. Fig. 5.9 shows failed constructive attempts to prove $\neg(\neg E \wedge \neg F) \vdash E \vee F$, the other half of the equivalence proved in Fig. 5.7. In a constructive proof all that you can do with a disjunctive conclusion is to throw half of it away with $\vee$ intro (5.9(a)) and all you can do with a negated premise is to derive a contradiction with $\neg$ elim (5.9(c)). In this example reducing uncertainty in the conclusion with $\vee$ intro doesn't help: the premise has something to do with both E and F, and the proof can't focus on just one of the two. Searching for a contradiction in the premise seems absurd: it only says 'suppose you don't have $\neg E$ and $\neg F$ together'.

If you don't notice the absurdity early enough you can push the symbols till you reach a point where you are trying to show that from F you can prove E (5.9(f)). The only hope of a proof would be to prove a contradiction, but it's certain that you don't have $\neg F$, because you already have $\neg E$ on line 2 and the premise says you can't have both together. Surely it's obvious that this is a dead end.

In fact we need a classical proof: De Morgan's Laws are classical equivalences. Classical proof needs the classical contra rule. The principles of using classical contra are

1. you have to use it backwards like an intro rule;
2. you match an open conclusion to the **consequent** A of the rule.

Whereas every other intro rule has a consequent which will only match a formula with an appropriate principal operator, the consequent A of classical contra will match **any** conclusion formula at all. That means that you can always make a classical contradiction step backwards and you never have a clue, either from the shape of the conclusion you are trying to prove or the hypotheses you are trying to prove it from, just when to do it. That's enough to prompt a very condemnatory slogan.

Table 5.4 classical contra slogan

Classical contradiction is hard to deal with; use it **only when you have to**.

I don't know any useful rules of thumb about when to use classical contradiction, except to say that you use it when you are stuck, but not usually at the point where you get stuck. Instead, you usually backtrack and try it earlier. As for how far to backtrack — experience and reflection is the only way to find out.

Fig. 5.10, for example, shows a classical proof of $\neg(\neg E \wedge \neg F) \vdash E \vee F$. Once you have explored the impossibility of making a constructive proof it's clear that there is nothing useful which you can do with the conclusion or the hypothesis. The only alternative is to start with a classical contradiction (5.10(a)). Then you have a choice: $\neg$ elim forward can be used on the premise or the new assumption. Because $\wedge$ is easier to work with in a conclusion than $\vee$, I choose the premise, giving 5.10(b). Backward steps using the slogans get us to 5.10(c), at which point the only available steps are $\neg$ elim with the premise or the assumption. Trying the assumption (we already used the premise) gives 5.10(d); then it's obviously

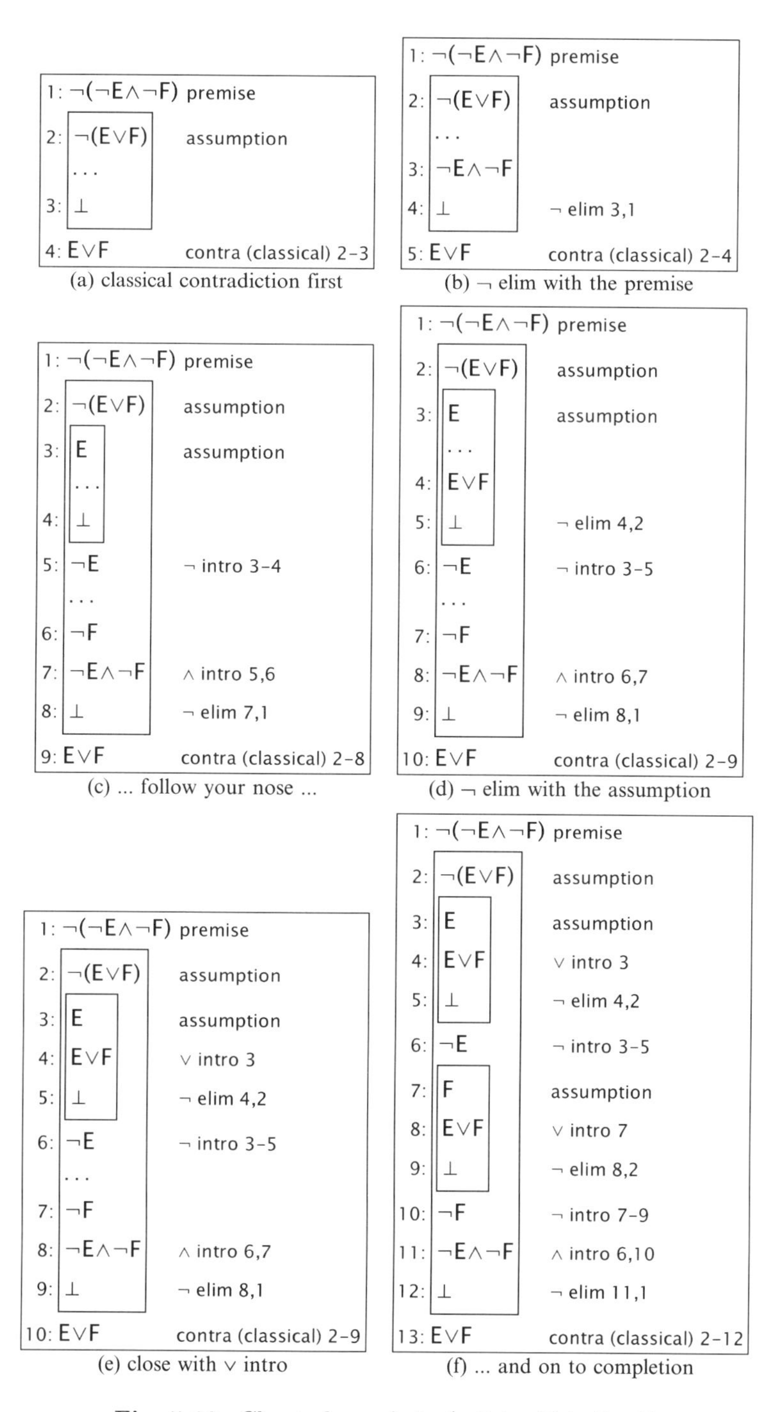

(a) classical contradiction first

(b) ¬ elim with the premise

(c) ... follow your nose ...

(d) ¬ elim with the assumption

(e) close with ∨ intro

(f) ... and on to completion

Fig. 5.10 Classical proof of $\neg(\neg E \wedge \neg F) \vdash E \vee F$

late enough for $\vee$ intro, closing that arm of the proof and giving 5.10(e). The other arm works in a similar way. The end result in 5.10(f) is a classical proof because of the classical contra step on line 13.

The proof looks very surprising if you forget the search process and read Fig. 5.10(f), as most people would, 'forwards', top-to-bottom. What, for example, inspired the proof designer, apparently from out of nowhere, to assume $\neg(E \vee F)$ on line 2? When you have made the proof search you know that the need for an assumption arose naturally from the need to make a classical contradiction on line 13 when nothing else would do, that $\neg(E \vee F)$ was just the assumption that the rule threw up, and that despite what the proof listing appears to suggest line 13 was the first step that was decided on. Imagine how difficult it must be to make proofs forwards-only, when you have to try to make assumptions without knowing what assumptions to make! Even deciding when to make an assumption, in such a cock-eyed scheme, would seem difficult.

By contrast, constructive proof search is straightforward. Even when classical contradiction has to be used, proof search is still highly rational. You'll find proof search with the rules of Chapter 3, up to and including classical contradiction, so easy that a few hours practice with Jape and its list of conjectures will make you an expert.

5.6.7 The law of excluded middle and the classical contra dance. $E \vee \neg E$ doesn't have a constructive proof. The only rule applicable by slogan is $\vee$ intro, and that asks us to prove either E or $\neg E$ from no premises at all. I'm sure you can see that proving E from no premises isn't possible — ought not to be possible, surely! — and that the same applies to $\neg E$.

The 'law' is dear to the classical heart, though. We ought to be able to prove it, and from what you know already you should realize how: start with a classical contradiction step (Fig. 5.11(a)). The next step just has to be $\neg$ elim — it's the only connective that's available — and the result, Fig. 5.11(b), shows the weird classical contradiction dance: if you can't prove a conclusion, then try proving it from its own negation! Classical contra wraps the conclusion in a negation and installs it as an assumption from which you must prove a contradiction; $\neg$ elim on the new assumption gives you the conclusion back. Weird, or what?

Now it's safe to use $\vee$ intro to throw half the conclusion away, because the other half is still preserved in the assumption. I keep $\neg E$ (5.11(c)), because it has a connective and that could be useful. Then $\neg$ intro gets us back to having to prove a contradiction (5.11(d)), and all we can do is use $\neg$ elim again with the original assumption. The proof closes with $\vee$ intro.

There are lots of pleasing symmetries in the proof search: for example, $\neg$ elim followed by throwing away half the conclusion is used twice, once with each half. Read forward the proof seems quite marvellous, but apart from the

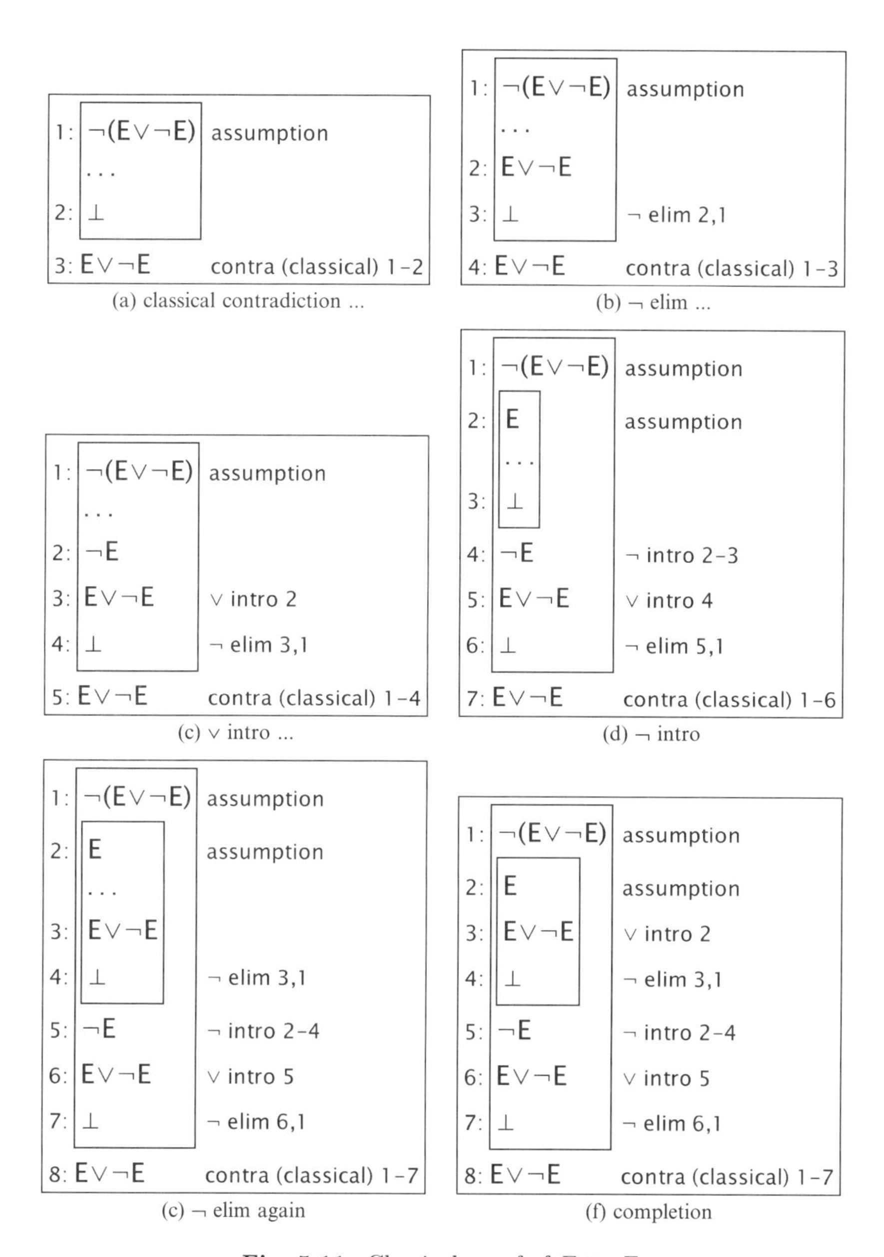

(a) classical contradiction ...

(b) ¬ elim ...

(c) ∨ intro ...

(d) ¬ intro

(c) ¬ elim again

(f) completion

Fig. 5.11 Classical proof of $E \vee \neg E$

first step it was an entirely connective-driven business. It was also built almost entirely backwards, apart from the ¬ elim steps that exploit the assumption of the classical contradiction step.

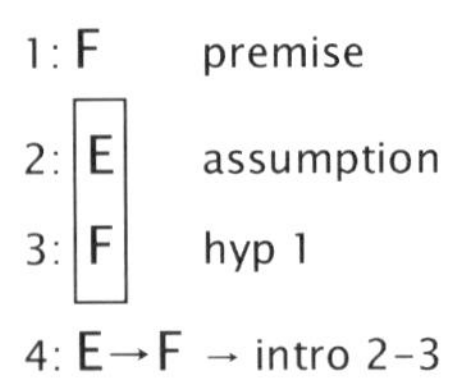

Fig. 5.12 Proof of the price-of-tomatoes principle, with `hyp` step

5.7 Just one more thing: redirection with `hyp`

The formal proof in Jape of the price-of-tomatoes argument $F \vdash E \to F$ (Fig. 5.12) uses a proof step that I haven't described before. We have a proof of F inside the box because there's a proof of F outside it. We have to write F on line 3 as the conclusion of the box because that's what the $\to$ intro step, which called for the box, demands. We could repeat the proof of F inside the box (in this case, merely writing "`premise`" again), but in general it makes sense to write something next to that conclusion line which says "look, there's already a proof over there". Jape calls this redirection "`hyp`", and it's no more than a means of avoiding repetition in box-and-line proofs. (It doesn't have a proof-tree equivalent because it would turn trees into DAGs — that's why it doesn't appear in Chapter 3.)

Jape puts `hyp` steps in automatically. On paper and on the blackboard you have to put them in for yourself. They don't add anything to the power of Natural Deduction, but they avoid a little repetition and they do no harm.

6 The logical quantifiers

So far we have been able to deal with logical claims about particular things: claims about somebody called Richard, claims about the name Richard, claims about Richard being a cabbage and/or a Martian. We haven't yet been able to relate claims about generalities to claims about particulars. From the claims 'all those integers are odd' and 'integer i is one of them', for example, we ought to be able to deduce 'integer i must be odd'. From the claim 'there is no Santa Claus' we ought to be able to deduce 'you're not Santa Claus'. From the claims 'that is a wild wolf' and 'we are somewhere in England' we ought to be able to deduce 'there is at least one wild wolf in England'.

Natural Deduction deals with generalizations, specializations, search for an example, and all the other ways that a claim can be about a collection of things or about a thing chosen from a collection, by using **quantified** formulae. In this book there are only two **quantifiers**:

Table 6.1 Quantifiers of Natural Deduction

Quantifier	Simple name	Latinate name
$\forall$	For all	Universal
$\exists$	There exists	Existential

Roughly, $\forall x(P(x))$ — pronounced 'for all x, $P\ x$' — means "every thing has property P", and $\exists x(P(x))$ — pronounced 'there exists x such that $P\ x$' — means "some thing has property P". But, as you saw with the connectives, logic isn't content with rough definitions. In order to reason, we must be precise and fundamental and, especially with the quantifiers, we have to avoid falling into paradox-traps.

6.1 A logical universe

To make a generalization we have to say what we are generalizing over: the streets of your home town, the positive integers less than 1000, all the dogs in Australia, all the sheep in the world, whatever. The collection we generalize over is our **logical universe**, and we can make claims about the whole universe, or some sub-universe, or about particular **individuals** in the universe.

To describe a universe is to point to some collection/pile/group/huddle of things and say "there are all the individuals that I'm considering". Names describe individuals in the universe, but a name is just a label: you can't deduce anything from a name. The same individual can even have several names: "2", "two", "the smallest prime" and "the positive square root of 4", for example, are all different names of the same number.

To begin with, while I'm setting up the ideas that lie behind the logical treatment of generalization, specialization and all the rest of it, I'm going to be vague about the universe I'm considering. It will be just some collection of things I can name, including abstract ideas, scientific theories, particular numbers, individual lumps of mud, people, football players, logicians, wolves — whatever suits my purposes.

If I tried to be precise about such an unconstrained universe I would risk falling into the set-theoretic hole that Russell pointed out to Frege (see Chapter 1). When I get more precise I'll restrict myself to simple well-understood things I can point to, like babies, wild wolves and finite integers, to avoid that paradox.

6.2 Properties of individuals

Consider the music-stave mnemonic "Every Good Baby Deserves Favour". Suppose there is some individual in the universe named i (not necessarily a baby: it might be a lump of mud, or a wild wolf, or Newton's third law of motion). I'm going to invent some properties which the individual i might have or might lack: the property of being a baby, the property of being good, the property of deserving something. My desk has perhaps one of these properties; each of my grandchildren has at some time had all three; the logic we are studying has one of them and so, occasionally, does my wastebasket.

Suppose that $\mathsf{Baby}(i)$ — pronounced "Baby i" — means "individual i has the Baby property". Individuals that "have the Baby property" are babies, of course, so $\mathsf{Baby}(i)$ is just the way to say that i is a baby. Similarly, $\mathsf{Good}(i)$ is supposed to mean i is good.

Then $\mathsf{Baby}(i) \wedge \mathsf{Good}(i)$ — or, equivalently, $\mathsf{Good}(i) \wedge \mathsf{Baby}(i)$ — says that i is a good baby: if you accept it, then you accept that i is both a baby and good; you can reject it only if i is not a baby or is bad. Suppose that $\mathsf{Deserves}(x, y)$ means "individual x deserves individual y": then $\mathsf{Deserves}(i, \mathit{favour})$ means that the i individual deserves the *favour* individual.

Now that I've explained my notation, I can write down a formula which means "if i is a good baby then i deserves *favour*":

$$\mathsf{Baby}(i) \wedge \mathsf{Good}(i) \rightarrow \mathsf{Deserves}(i, \mathit{favour}) \tag{6.1}$$

It's easy to see that if there are other individuals in the universe then I can replace i with their name and produce a similar claim, but with different impact:

$$\mathsf{Baby}(richard) \wedge \mathsf{Good}(richard) \rightarrow \mathsf{Deserves}(richard, favour) \quad (6.2)$$

If I'm the individual *richard* then this claim doesn't say that I deserve favour, because it's an implication and I'm not a baby: it's a cunning-uncle promise. It doesn't say that I don't deserve favour either, only that I would certainly deserve it if I was ever a good baby again.

6.3 Generalization and specialization

I can make a claim about the i individual, stating some of its properties and its relationship to the *favour* individual. But the claim I started with was a generalization, a claim about every good baby and its deserts. I **generalize** my claim about the individual i by crossing out the name i wherever it occurs and replacing it with a variable name x. Then I wrap the result up as a **universal quantification**, written $\forall x(\ldots)$, and I have a claim which applies to any individual whose name you write in place of x.

$$\forall x(\mathsf{Baby}(x) \wedge \mathsf{Good}(x) \rightarrow \mathsf{Deserves}(x, favour)) \quad (6.3)$$

is pronounced "for all x, if x is a baby and x is good, then x deserves *favour*". It means precisely that every good baby in the universe deserves favour.

The reverse of generalization is **specialization**. A universal claim applies to **every** individual, so we can replace the variable with the name of **any** individual, get rid of the quantifier, and there we are. Here are four specializations of (6.3):

$$\mathsf{Baby}(i) \wedge \mathsf{Good}(i) \rightarrow \mathsf{Deserves}(i, favour) \quad (6.4)$$
$$\mathsf{Baby}(j) \wedge \mathsf{Good}(j) \rightarrow \mathsf{Deserves}(j, favour) \quad (6.5)$$
$$\mathsf{Baby}(richard) \wedge \mathsf{Good}(richard) \rightarrow \mathsf{Deserves}(richard, favour) \quad (6.6)$$
$$\mathsf{Baby}(favour) \wedge \mathsf{Good}(favour) \rightarrow \mathsf{Deserves}(favour, favour) \quad (6.7)$$

The first and the third we've seen already. The second is just like the first, a claim about the properties of individual j. The last looks a bit odd: *richard* was once a baby, but *favour* is surely a different kind of thing. Shouldn't that claim be outlawed somehow?

The answer's no: it shouldn't be outlawed. If *favour* is an individual in the universe, as I'm supposing it is, then it can be examined for properties too. That's what the left-hand side of the implication does. This claim is just exactly as vacuous as the one about *richard*, and for exactly the same reason: whatever sort of thing *favour* is, it isn't a baby, so the claim is in cunning-uncle territory and therefore useless.

Universal quantifications are universal claims, claims about every individual in the universe. You can't outlaw anybody or anything, once you've decided they're in the universe to begin with.

6.3.1 Small print. When I generalized (6.1) I crossed out **all** the is and replaced them with xs. I didn't have to do that: it's enough to cross out and replace **some** of the occurrences of a name. $\forall x(\mathsf{Baby}(x) \wedge \mathsf{Good}(i) \rightarrow \mathsf{Deserves}(x, \mathit{favour}))$, for example, says that if i is good then every baby in the universe deserves favour. I can even cross out **none** of the occurrences and make a vacuous generalization: $\forall x(\mathsf{Baby}(i) \wedge \mathsf{Good}(i) \rightarrow \mathsf{Deserves}(i, \mathit{favour}))$ always specializes to (6.1).

When specializing a universal claim, on the other hand, I can choose the individual freely but I must replace **all** the occurrences of the quantified variable name with the chosen individual's name.

6.4 Anonymization and nomination

When we are feeling specially sugary, we might accept that i actually is a good baby who deserves favour (no ifs and no buts — so no implication, just a conjunction):

$$\mathsf{Baby}(i) \wedge \mathsf{Good}(i) \wedge \mathsf{Deserves}(i, \mathit{favour}) \tag{6.8}$$

That is a specific claim, identifying a specific favour-deserving individual named i. We can **anonymize** it, make it say that there is some exceptional individual out there, but avoid naming that individual. As before we replace occurrences of a name (i) with a variable name x, but this time we wrap up the result in an **existential quantification**, written $\exists x(\ldots)$. Now we have a claim about a particular unnamed individual in the universe:

$$\exists x(\mathsf{Baby}(x) \wedge \mathsf{Good}(x) \wedge \mathsf{Deserves}(x, \mathit{favour})) \tag{6.9}$$

This is pronounced "there exists an x such that x is a baby and x is good and x deserves *favour*", and it means that there is a good deserving to-be-favoured baby somewhere out there in the universe. Since we already accept that i is such a baby, we must accept the anonymized claim.

The reverse of anonymization is **nomination**. We name a particular individual as a **witness** to demonstrate an existence formula. In this case I could use i, because that is the example I anonymized to begin with. But I don't have to: if j is another equally worthy baby then j can be my witness instead.

$$\mathsf{Baby}(j) \wedge good(j) \wedge \mathsf{Deserves}(j, \mathit{favour}) \tag{6.10}$$

If I were to attempt to nominate the *favour* individual then I would produce an unacceptable claim:

$$\mathsf{Baby}(\mathit{favour}) \wedge \mathsf{Good}(\mathit{favour}) \wedge \mathsf{Deserves}(\mathit{favour}, \mathit{favour}) \qquad (6.11)$$

Not so! There's no implication, so no chance of a cunning-uncle trick here: *favour* is not a baby, so it doesn't support the existential claim. We can't nominate any old individual to witness an existential claim. We have to find one which **satisfies** the claim: that is, one that makes the claim provable (in constructive logic) or true (in classical logic).

6.4.1 More small print. Just as with universal claims in Section 6.3.1, so it is with existentials. When anonymizing I cross out some or all or even none of the occurrences of a chosen name (i in the example above); when nominating I must replace all the occurrences of the quantified variable (x in the example above) with a chosen name.

6.5 Predicates and relations: formulae with holes

Goodness and babyhood are properties which an individual can have; $\mathsf{Good}(i)$ and $\mathsf{Baby}(i)$ are formulae which say that the individual i has those properties. Abstracting, $\mathsf{Good}(\_)$ and $\mathsf{Baby}(\_)$ — the same formulae but with holes in place of the name i — are templates for constructing remarks about arbitrary individuals, **simple predicates** which describe a particular property.

We aren't restricted to simple predicates: a predicate is, in general, just a formula with some number of holes in it. $\mathsf{Good}(i)$ is a simple claim about an individual i; $\mathsf{Baby}(i) \wedge \mathsf{Good}(i)$ a more complicated claim; $\mathsf{Baby}(i) \wedge \mathsf{Good}(i) \rightarrow \mathsf{Deserves}(i, \mathit{favour})$ a more complicated claim still. If we cross out all the is we get

$$\mathsf{Baby}(\_) \wedge \mathsf{Good}(\_) \rightarrow \mathsf{Deserves}(\_, \mathit{favour}) \qquad (6.12)$$

This is a **composite predicate**: more than just a simple name and a single hole, but a predicate still. If we call this formula $\mathsf{BGDfavour}(\_)$, then $\mathsf{BGDfavour}(i)$ is the claim I started with in (6.1); $\mathsf{BGDfavour}(j)$ is (6.5); $\mathsf{BGDfavour}(\mathit{favour})$ is (6.7), and so on.

In making a composite predicate you don't have to be straightforward: you can play tricks. The composite predicate $\mathsf{BGDifavour}(\_)$, produced by deleting the first two occurrences of i but leaving the third, is

$$\mathsf{Baby}(\_) \wedge \mathsf{Good}(\_) \rightarrow \mathsf{Deserves}(i, \mathit{favour}) \qquad (6.13)$$

$\mathsf{BGDifavour}(j)$ claims that if j is a good baby then i deserves favour. (Everybody, at some stage of their life, thinks the world is treating them like j and giving everything to some i.)

You can even play the trick of leaving no holes at all. For all sorts of reasons (see the discussion of zero and the green sheep below) mathematicians don't want to make zero a special case. So a predicate formula with no holes is just a very peculiar predicate: every instance is the same as every other. Such a predicate is as tricky as a cunning uncle, and just as hard to legislate against.

Composite predicates can be arbitrarily complicated formulae. The claim that every good baby deserves favour — $\forall x(\mathsf{Baby}(x) \wedge \mathsf{Good}(x) \rightarrow \mathsf{Deserves}(x, \mathit{favour}))$ — can be made into a predicate by crossing out *favour*. That produces a composite predicate which is a universal quantification with a hole in it:

$$\forall x(\mathsf{Baby}(x) \wedge \mathsf{Good}(x) \rightarrow \mathsf{Deserves}(x, \_)) \tag{6.14}$$

This predicate, which we might call EBGD, can now be applied to individuals to produce claims. $\mathsf{EBGD}(\mathit{favour})$ — the predicate with *favour* in the hole — is the claim we've been working with. We might want to say $\mathsf{EBGD}(\mathit{icecream})$, a claim which will gain the approval of most babies but fewer parents.

If you cross out occurrences of more than one name then you produce a **relation**, a formula which describes how one individual relates to another. To keep things straight you have to mark the holes to show which name went where. $\mathsf{EBGDr}(\_1, \_2)$, for example, might describe the formula

$$\mathsf{Baby}(\_1) \wedge \mathsf{Good}(\_1) \rightarrow \mathsf{Deserves}(\_1, \_2) \tag{6.15}$$

— the is were in the $_1$ positions, *favour* was in the $_2$ position. $\mathsf{EBGDr}(i, \mathit{favour})$ expresses the relationship between good baby i and *favour* in (6.1). $\mathsf{EBGDr}(\mathit{richard}, \mathit{favour})$ reproduces the vacuous remark of (6.6).

6.5.1 Multiple quantifiers. The specialisation/nomination process I've described several times — cross out the xs, replace them by is, throw away the quantifier — works fine, no matter how complicated the composite predicate might be. In particular, it works even if the composite predicate contains quantifiers. For example, to nominate a witness for $\exists y(\forall x(\mathsf{Baby}(x) \wedge \mathsf{Good}(x) \rightarrow \mathsf{Deserves}(x, y)))$, you cross out the y, replace it with a witness like *icecream*, throw away the $\exists$, and produce $\forall x(\mathsf{Baby}(x) \wedge \mathsf{Good}(x) \rightarrow \mathsf{Deserves}(x, \mathit{icecream}))$.

Matters get more complicated if a formula contains quantifiers which share a variable. To avoid difficulty I shall strictly avoid such practices.

6.6 Matching quantified-formula schemes

Chapter 4 dealt with the matching of connective-formula schemes to formulae. Matching $A \rightarrow B$ to $\mathsf{Baby}(i) \wedge \mathsf{Good}(i) \rightarrow \mathsf{Deserves}(i, \mathit{favour})$, for example, matches A with $\mathsf{Baby}(i) \wedge \mathsf{Good}(i)$ and B with $\mathsf{Deserves}(i, \mathit{favour})$.

Table 6.2 Binding power of connectives and quantifiers

1. $\forall$, $\exists$ (right-to-left)
2. $\neg$ (right-to-left)
3. $\wedge$ (left-to-right)
4. $\vee$ (left-to-right)
5. $\rightarrow$ (right-to-left)

What happens, though, if we match the quantified-formula scheme $\forall y(Q(y))$ with $\forall x(\mathsf{Baby}(x) \wedge \mathsf{Good}(x) \rightarrow \mathsf{Deserves}(x, \mathit{favour}))$? Clearly y in the scheme should match x in the formula, and clearly $Q(y)$ has to match $\mathsf{Baby}(x) \wedge \mathsf{Good}(x) \rightarrow \mathsf{Deserves}(x, \mathit{favour})$, just because of the way that the brackets fall. But what matches Q?

The answer is that the scheme predicate $Q(\_)$ matches the composite predicate formula $\mathsf{Baby}(\_) \wedge \mathsf{Good}(\_) \rightarrow \mathsf{Deserves}(\_, \mathit{favour})$ — and then $Q(y)$, because y matched x, is $\mathsf{Baby}(x) \wedge \mathsf{Good}(x) \rightarrow \mathsf{Deserves}(x, \mathit{favour})$. It all fits.

When we don't have brackets to guide us, we need binding rules. Table 6.2 shows that quantifiers fit in at the highest priority, binding tighter than any connective. So, for example, $\neg\forall x(\mathsf{Good}(x))$ is a negation, equivalent to $\neg(\forall x(\mathsf{Good}(x)))$, and $\forall x(\mathsf{Good}(x)) \wedge \exists y(\mathsf{Baby}(y))$ is a conjunction, equivalent to $(\forall x(\mathsf{Good}(x))) \wedge (\exists y(\mathsf{Baby}(y)))$.

6.7 Universes, individuals, 'actual i'

So far I've based my discussion on a universe which was understood as 'every thing I can name'. I didn't consider the possibility that there might be no things which could be named. The universe of nameable things, simply understood, is not empty — it's always got me in it, for one, and you, for another.

There are lots of other non-empty universes: the universe of whole numbers, the universe of rational numbers, the universe of people who have won Olympic gold. On the other hand, there are universes that are empty of individuals: the universe of square primes (empty by definition); the universe of people who have won a gold medal at twenty successive Olympics (empty now, but who knows what hyperathlete may emerge?); the universe of wild wolves living in England (empty in 2004, but I dream that one day England will be fit for their return).

If the universe of quantification is populated then certain consequences follow. Suppose, for example, I accept $\forall x(\mathsf{Good}(x))$ and the universe isn't empty: then there must be at least one good individual in it. As a logical consequence of $\forall x(\mathsf{Good}(x))$ I must accept $\exists x(\mathsf{Good}(x))$ — some particular individual is good, whether or not I know its name.

But what happens if the universe of quantification is empty? What can we say, for example, if we choose the universe of wild English wolves? There are no wild wolves in England, so we can't say $\exists x(\mathsf{Good}(x))$, because we can't point to a good wild English wolf. But we also can't say $\exists x(\neg\mathsf{Good}(x))$, for an exactly similar reason.

You might suppose that we couldn't say anything definite about the inhabitants of an empty universe, but you'd be wrong, as we shall very soon see. Just like zero, the most famous kind of useful Nothing, empty universes aren't devoid of properties. In fact we can conclude $\forall x(\mathsf{Good}(x))$ of the universe of wild English wolves! But then $\exists x(\mathsf{Good}(x))$ can't be a logical consequence if our logic is to avoid nonsense.

So we could be misled in our reasoning if we assumed that every logical universe is populated. Logicians deal with this difficulty in various ways. I shall use a version of Natural Deduction in which you must always take special care to point to evidence about presence when specializing or nominating. A special kind of **presence marker** — actual i, actual j, and so on — will be used in the formal rules as this kind of evidence.

6.8 Reasoning with ∀

Conjunction can be used to string together a collection of assertions $A \wedge B \wedge C \wedge \ldots$, to say that we accept every single one of the assertions in the collection. If we want to make a conjunctive claim about every individual in a very large collection — $P(C_1) \wedge P(C_2) \wedge \ldots$ — it is often best to make a general claim about the collection as a universe of quantification. In this way we can deal even with infinite collections: for example, Goldbach's conjecture says something about every even number larger than 2.
The claim $\forall x(P(x))$ is pronounced "for all x, $P\ x$".

Definition 6.1 $\forall x(P(x))$ claims that every individual in the universe satisfies predicate $P(\_)$.

The definition doesn't say that you will ever meet an individual which satisfies predicate $P(\_)$, because the universe might be empty. Like the definition of implication, it's about what would happen if you did meet anybody. The predicate $P(\_)$ can be composite, so that $P(\_)$ is, in general, just a formula with some holes in it (or even no holes at all, if you want an irrelevant quantification).

Suppose that I accept $\forall x(P(x))$, and I accept also that there is an individual in the universe called i. Then I must accept that the individual i, since it is in the universe, has the property P. That is, I must accept $P(i)$. This is the step of **specialization**, captured by the ∀ elim rule in Table 6.3. You cross out all the

Table 6.3 Rules for universal quantification

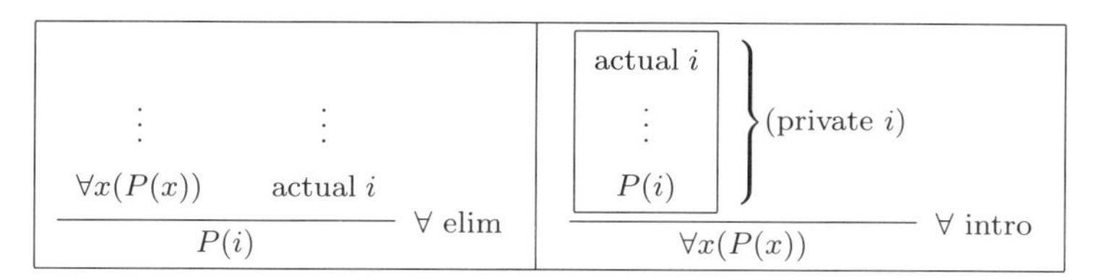

xs in $P(x)$, exposing the predicate $P(\_)$ as a formula with holes, and then write i in each of the holes, creating $P(i)$.

How might I be persuaded to accept $\forall x(P(x))$? In very small universes you might take me round to meet all the individuals in turn, showing me that each has property P. By showing me all of them, you force me to accept $\forall x(P(x))$.

But in many cases, and in most of the interesting cases, the universe isn't small at all. For example, how could you prove that in the infinite universe of integers, $\forall x(x > 1 \rightarrow x^2 > x)$? In such a case you have to make a **generalized proof** which would apply to any individual in the universe, no matter how I pick it out.

You begin by asking me to suppose that I've picked something from the universe. It mustn't matter which individual I pick: for example, if I'm picking from a universe of integers, it mustn't matter if it's positive, negative, zero, non-zero, prime, non-prime, a power of 2, a cube, ...whatever. I don't really pick anything (this isn't the magician's trick "think of a number"), I just imagine that I've picked something. 'For the purposes of argument' we name this imaginary individual i (or j or k or any other pseudo-name which isn't the real name of anything in the universe). We capture the imagination step with the presence marker 'actual i' (or actual j, or actual k, any other pseudo-name you choose); that means "suppose that there is some individual i (or j or k or ...) in the universe".

Now you must prove that the imagined individual must necessarily have property P — that is, from the assumption actual i, plus any premises and any stuff you have already proved from the premises, you must show me that I must accept $P(i)$. To mimic the notion that i is an arbitrary choice I have to be sure that you haven't already persuaded me to accept something about the individual i elsewhere, in some other part of the proof. We impose a special **side condition** that the name i must be **private** to the argument. It isn't a hard condition to meet — if you've used the name i before in your proof, just choose a different name like j or k or *foodle* that you haven't used. The privacy condition makes it impossible for you to smuggle in any extra assumptions about i from outside the 'i must have property P' argument, because you can't mention i anywhere outside that argument.

If you can prove, under conditions of privacy, that the assumption that i is an individual means that it must have property P, then you have reached a marvellous conclusion. Your argument, which seemed to be about an particular individual i was, because of its abstraction and its isolation, about an **arbitrary** individual.[1] If I now point to a real individual — *favour* or *richard* or *icecream* or whoever — the proof must apply to that real individual, because that might have been the one I chose. The same applies to every individual, so the proof applies to **any** individual in the universe: any individual I might meet will have property P. I must accept $\forall x(P(x))$!

Notice that the generalized proof strategy which is captured by $\forall$ intro doesn't require that the universe really does contains any individuals. You prove that **if** you meet an individual, **then** that individual will have property P. Even if you can separately prove that there are no individuals to meet, the proof of $\forall x(P(x))$ still stands. We'll see the consequences of that oddity very soon.

The generalized proof technique is captured in the $\forall$ intro rule in Table 6.3. The privacy condition is absolute. You can't mention i outside the antecedent proof **at all**. In particular, you can't mention it anywhere in the predicate $P(\_)$ — which is, remember, just $P(x)$ with the xs crossed out.

Since universal quantification is a generalization of conjunction, the $\forall$ elim rule is a generalization of $\wedge$ elim: we pick something from the long, possibly infinite conjunction $P(C_1) \wedge P(C_2) \wedge \ldots$, just as $\wedge$ elim picked one from a pair.

Similarly, the $\forall$ intro rule is a generalization of $\wedge$ intro. The generalized proof stands for a sequence of individual proofs $P(C_1)$, $P(C_2)$, ... and the consequence wraps up the whole sequence into a single short claim.

6.9 Reasoning with $\exists$

Disjunction can be used to string together a collection of assertions $A \vee B \vee C \vee \ldots$, to say that we accept one or more of the assertions in the collection. If we want to make a disjunctive claim about the individuals in a large collection — $P(C_1) \vee P(C_2) \vee \ldots$ — it is often best to make an existence claim about the collection. In this way we can talk even about infinite collections, as for example in the proof of Fig. 3.2 on page 38, where it was asserted that irrational numbers x and y exist such that x^y is rational.
The claim $\exists x(P(x))$ is pronounced "there exists x such that P x".

Definition 6.2 $\exists x(P(x))$ claims that I can point to an individual which satisfies predicate $P(\_)$.

[1] Often referred to as an **arbitrary fixed** individual, because it is arbitrarily selected but that selection is fixed during the argument.

The definition doesn't force me to point to any individual: like the definition of disjunction, I'm allowed to leave the question uncertain. As usual $P(\_)$ can be a composite predicate, in general just a formula with some holes in it, or (if you want an irrelevant quantification) even no holes at all.

$\exists$ is a generalization of $\vee$, so we'd expect to use an acceptance of $\exists x(P(x))$ in a generalized argument by cases. Suppose that I accept $\exists x(P(x))$: how can you persuade me that I must accept some consequence C? You know that I accept that there is some individual which has the property P, but you know that I won't say which. The $\vee$ elim rule solves the problem by dealing with each case of the disjunction separately, but you can't really expect to deal with all the cases in the generalized disjunction, unless the universe is finite and rather small.

So you ask me to suppose that there is something called i which actually has property P, and then show me that in those circumstances I must accept C. Just as in the case of $\forall$ intro, provided that you don't make any prior assumptions about the individual I imagine, your proof applies to any individual which happens to have property P. If there really is an individual out there with property P — which is what my acceptance of $\exists x(P(x))$ means — I must accept C. You have done your job without forcing me to reveal any individual that really has property P.

That anonymized proof strategy is captured in the $\exists$ elim rule of Table 6.4. Once again, the privacy condition is absolute: in particular i can't appear in the conclusion C (because C appears outside as well as inside the private proof) or in the predicate $P(\_)$.

How might I be persuaded to accept $\exists x(P(x))$, if I don't accept it already? That's easy: point to an individual which actually has property P, and use it as a **witness** to the existence assertion. That's captured in the $\exists$ intro rule of Table 6.4.

Existential quantification is a generalization of disjunction, so $\exists$ elim is a generalization of $\vee$ intro: the general proof stands for all the separate little proofs $P(C_1)$, $P(C_2)$, ...

Table 6.4 Rules for existential quantification

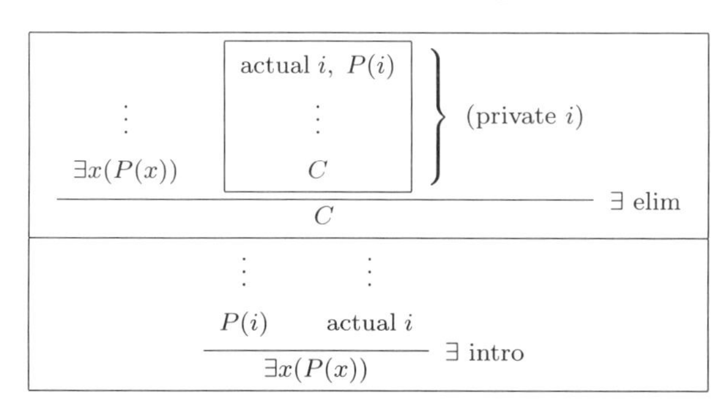

Similarly, $\exists$ intro is a generalization of $\vee$ intro: show that I must accept one of the components of the large, possibly infinite, disjunction and the whole claim follows.

6.10 Quantifier idioms

The connectives of Chapter 3 are as simple to use as the familiar operators of arithmetic. Quantifiers are a little more subtle.

6.10.1 $\forall x(P(x) \rightarrow Q(x))$**.** At the beginning of this chapter I said I would deal with a universe that includes everything you can name. But the claim 'every good baby deserves favour' sounds like a quantification over a universe of good babies: any individual in that universe, it seems to say, deserves favour. If I had said that I would quantify over the good-baby universe, then all I would have had to say is $\forall x(\mathsf{Deserves}(x, favour))$. Instead I quantified over a universe that included many individuals that aren't good or aren't babies or both, and produced the formula $\forall x(\mathsf{Good}(x) \wedge \mathsf{Baby}(x) \rightarrow \mathsf{Deserves}(x, favour))$.

The idiom $\forall x(P(x) \rightarrow Q(x))$ is a nested promise: it promises that **if** you meet an individual, **then**: **if** that individual has property P, **then** it will also have property Q. That neatly picks out the sub-universe of individuals with property P, and says that they all have property Q.

6.10.2 $\exists x(P(x) \wedge Q(x))$**.** In the good-baby universe, the claim that some good baby out there really deserves an ice cream is just $\exists x(\mathsf{Deserves}(x, icecream))$. In the everything-you-can-name universe the same claim is $\exists x(\mathsf{Good}(x) \wedge \mathsf{Baby}(x) \wedge \mathsf{Deserves}(x, icecream))$. The more complicated formula makes the claim that I can point to an individual which is, because of the conjunction, **both** a good baby **and** deserving of an ice cream. That's an individual which is in the universe of good babies and in the universe of those who deserve an ice cream.

The formula $\exists x(P(x) \wedge Q(x))$ is a pointing claim — I can point to an individual which has property P and property Q. You can read either P or Q as defining a universe, since conjunctions are symmetrical. So the claim $\exists x(\mathsf{Good}(x) \wedge \mathsf{Baby}(x) \wedge \mathsf{Deserves}(x, icecream))$ captures each of the claims that there is somebody in the good-baby universe who deserves an ice cream, or that there is somebody in the deserves-an-ice-cream universe who is a good baby, or that there is an individual in the good universe who is a baby deserving an ice cream, or that there is somebody in the baby universe who is good and ice-cream-deserving.

Usually the idiom $\exists x(P(x) \wedge Q(x))$ is read from left to right, expressing the claim that P defines a sub-universe of individuals, one or more of whom has property Q.

6.11 Universal quantification and the empty universe

Zero was once a revolutionary idea. First of all it was just a brilliant wheeze: accountants and the like saw the need for a mark which means Nothing. Better to make a mark than to leave no mark, because absence is ambiguous: it might mean there is Nothing here, or it might mean there is nothing here. Once you have a mark which means Nothing, absence of a mark is distinct from a mark of absence.

The idea of a name which means 'nothing' is easily mocked. Lewis Carroll, a 19th century logician, did that in *Through the Looking Glass*:

> "Look down the road and tell me who you can see", said the King.
> "I can see nobody on the road", said Alice.
> "What good eyesight you have!", exclaimed the King. "To see Nobody at such a distance! I have never seen him at all!"

The Babylonians knew something about zero, but the Greeks and Romans apparently did not. Our use of zero derives from Indian astronomers, who seem to have picked it up in about 400 CE from Hindu accountants in North India, who inherited or re-invented it in about 100 CE. Muhammad ibn Musa al-Khwarizmi (c. 800 CE), whose name was corrupted into Latin as 'Algoritmi' and thus gave us the word 'algorithm', learnt of the idea and incorporated it into his wonderful system for doing arithmetic, which every schoolchild nowadays learns (see Chapter 14). Europeans were slower off the mark — at first they called al-Khwarizmi's notation 'the nine signs', despite the obvious fact that there are ten signs including zero — but they caught up eventually and abandoned Roman stick-counting for what they still call Arabic numerals.

The properties of zero were developed over time as mathematicians incorporated it into their thinking. It isn't a counting number (you can see one sheep or two sheep, but you can't actually see zero sheep ...) so most of the world thinks of zero as a special case, different from other numbers if it's a number at all. Mathematicians, who love to generalize, try to make zero just another number, albeit a remarkable one.

For example: what is the value of $\sum\langle\rangle$, the sum of an zero-length (empty) sequence? Well, counting upwards we know that $\sum\langle a_0\rangle = a_0$, $\sum\langle a_0, a_1\rangle = a_0 + a_1 = (\sum\langle a_0\rangle) + a_1$. Generalizing, we expect that

$$\sum\langle a_0, a_1, \ldots, a_{n-1}, a_n\rangle = \left(\sum\langle a_0, a_1, \ldots, a_{n-1}\rangle\right) + a_n$$

If the zero-length sequence is to fit the generalization, we must have

$$\sum\langle a_0\rangle = \left(\sum\langle\rangle\right) + a_0$$

and since we already know $\sum\langle a_0\rangle = a_0$, that tells us that $\sum\langle\rangle = 0$. The only answer that fits is that the sum of a zero-length sequence, one in which there are no elements, is zero. Not undefined, not a stupid question, but zero.

Similar reasoning tells us that the value of the product of an empty sequence $\prod\langle\rangle = 1$ — surely even more surprising! — and that $x^0 = 1$, and so on and on. Zero isn't a ludicrous nothing-with-a-name, it's a number. Of course it's an unusual number: for one thing, it has as a factor any other number at all; for another, it can sometimes bite, as when $x \div 0$ is infinite (and infinity is another remarkable number, but outside this discussion).

Only the most naive schoolchild would nowadays think zero is just a trick, or not really a number. It is a number, and you can meaningfully do calculations with it which can have surprising results.

Unfortunately, schoolchildren don't have much experience with the empty set, which is another kind of nothing. If they did, novice logicians wouldn't have so much trouble with the field of green sheep, the room of drunken circus elephants, the class of attentive students (only joking!) or the pack of wild English wolves.

6.11.1 The field of green sheep. Suppose that I rig up a huge green light filter above a field of sheep, so that anybody who looks into the field will see green sheep. Taking the animals in that field as our universe, you'd surely agree that $\forall x(\mathsf{Green}(x))$ is a reasonable claim, provided that you're prepared to accept for the sake of argument that everything that looks green **is** green.

Now I send in a sheepdog with instructions to drive the sheep into another field. We no longer have $\forall x(\mathsf{Green}(x))$, because the dog looks black-and-green, but surely $\forall x(\mathsf{Sheep}(x) \rightarrow \mathsf{Green}(x))$ holds. Then, as each sheep leaves the animals-in-that-field universe, all the sheep that are left behind are still green. At the end, with only the dog in the field, do we have $\forall x(\mathsf{Sheep}(x) \rightarrow \mathsf{Green}(x))$? Well, the way that I built it, I think that we do: **if** you meet a sheep there, **then** it will seem to be green.

Then I turn the light off. It's still true that if you meet a sheep in this field it will seem to be green, because the dog is keeping them all out. I close the gate and call the dog off: now the field is completely empty and has no funny lighting but we still have $\forall x(\mathsf{Sheep}(x) \rightarrow \mathsf{Green}(x))$. All the sheep in an empty field are green (also red, blue, purple and polka-dot), and so are the dogs. The empty universe is a peculiar place.

We don't need an empty field to hold the green sheep: let me take you to the pen where I keep my puppy dogs. They aren't very well trained yet, and they make such a racket that all sheep keep well clear. In this universe, then, $\forall x(\neg\mathsf{Sheep}(x))$. But here too we have $\forall x(\mathsf{Sheep}(x) \rightarrow \mathsf{Green}(x))$: there are puppies but no sheep, so if you met a sheep then I can say it would be green, and you can't show me one that isn't. The argument is by vacuous implication, just like the cunning uncle of Chapter 3, and the formal argument is shown in Fig. 7.14 on page 115. The empty sub-universe is just as weird as the empty universe.

It turns out that the empty universe can be regarded as the root of any universe you like. I am fairly sure that in the room you are in, all the circus elephants are drunk; in England, as I write, all the wild wolves are good and at the same time all bad, all the female wolves are called Demelza, and all the males can whistle Beethoven's Ode to Joy in three-part harmony.

You can reject these kind of arguments if you wish, but you will be taking a position just like that of the mediaeval Europeans faced with the idea of zero, cutting yourself off from all sorts of useful reasoning. If you accept them then you can move forward, realizing that the empty universe is an interesting and surprising place, the zero of quantification, just as useful and dangerous as the number zero. No wonder we need presence markers in our rules to protect ourselves against it.

6.12 Quantifier rules summarized

Table 6.5 shows all the quantifier rules. There are only four of them. Together with the ten rules of Table 3.9 and two rules from Table 3.10, they make up the whole of the system of Natural Deduction. Add the `hyp` step of Section 5.7, and you have all that you need to make formal proofs in Jape. Sixteen rules, the whole caboodle. Hardly more than a page. Truly this is a **simple** formal system.

6.12.1 That's it! No more rules. Nothing more to be said. Connectives and quantifiers — done the lot.

6.12.2 Oh no it isn't! Of course there is more. What comes next is **use, practice and reflection**. Jape will help you to make formal proofs with quantifiers and connectives in any combination. It's tricky at first to use the rules with perfect accuracy; Jape does the accuracy bit, so you can concentrate on the steering.

Table 6.5 The quantifier rules

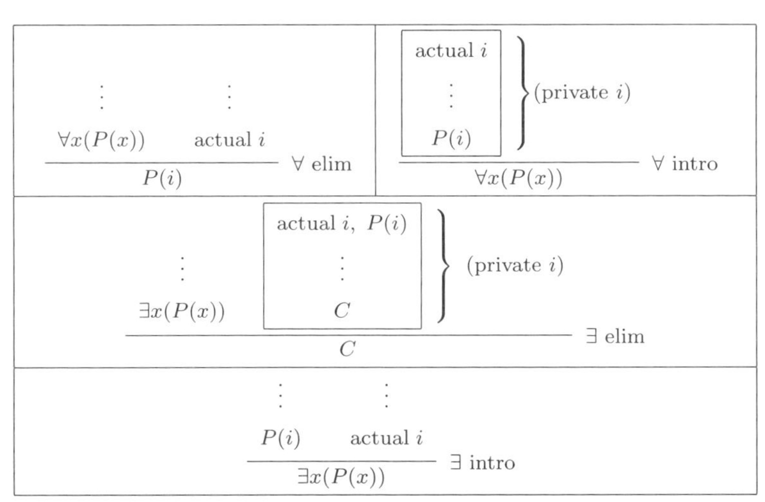

That's how you get practice. Reflection means turning over what you think you know to see if you understand it. Once you get some practice in, there will be plenty to think about!

And then, once you are a skilled prover, comes disproof in Part III and program proof in Part IV.

7 Proofs with quantifiers

Now that we have quantifiers, we can examine more interesting proofs. As in Chapter 5, I begin with some informal examples to give a feel of how the rules work.

7.1 The man in the dock

A man is on trial for murder. He's a minor gangster and between you, me and the judge, we think he did it. But English justice, like constructive logic, doesn't search for truth. Court cases are decided by argument; the law is about proof and disproof. (An oversimplification: for the rest of this chapter let's pretend.)

To get off, all our man has to do is to undermine the prosecution's argument, to raise a reasonable doubt in the mind of the court that he might be guilty. In order to do that, he and his lawyer have to show that there is a gap in the reasoning that leads from the prosecution's evidence to a guilty judgement and/or that their evidence is self-contradictory and/or that it contradicts reality. Because he once did a course on logic, he's going full tilt for contradiction. Because he didn't study very hard, he is on shaky ground.

7.1.1 The prosecution case. The prosecution brings a single witness — call him *enemy* — who testifies that he actually saw our man — call him *accused* — kill the victim. He identifies our man in court. You've seen the TV dramas; you can imagine the scene.

In earlier sessions, beyond the scope of this discussion, the court has heard evidence of the death, and is convinced that a death took place — in our notation, 'actual *killing*'. The defence doesn't dispute this: certainly somebody died. But it doesn't admit that our man committed murder.

What has the prosecution proved? Ignoring the problem of veracity — *enemy* might be lying or mistaken, which is one of our man's potential defences — it has shown somebody who says they saw the killing, and the court heard that person identify our man as the killer. The claims of the evidence can be boiled down to $\mathsf{Saw}(enemy, killing)$ — the enemy saw the killing — and $\mathsf{Identifies}(enemy, accused)$ — the enemy pointed to our man as the one he saw doing it. (Actually this is a long way from proof of murder. In English law, a crime requires a guilty mind. If you kill somebody accidentally, it may be a crime but it isn't murder. Still, let's pretend.)

Table 7.1 The 'legal axiom'

$$\forall x \left(\begin{array}{l} \mathsf{Person}(x) \rightarrow \\ \forall y \left\{ \begin{array}{l} \mathsf{Crime}(y) \rightarrow \\ \quad \exists z \ \langle \mathsf{Person}(z) \wedge \mathsf{Saw}(z, y) \ \wedge \mathsf{Identifies}(z, x) \rangle \rightarrow \mathsf{Guilty}(x, y) \end{array} \right\} \end{array} \right)$$

1.	actual *enemy*	we can see him
2.	$\mathsf{Person}(enemy)$	we can see that too
3.	$\mathsf{Saw}(enemy, killing)$	*enemy* claims
4.	$\mathsf{Identifies}(enemy, accused)$	*enemy* claims
5.	$\mathsf{Person}(enemy) \wedge \mathsf{Saw}(enemy, killing)$	$\wedge$ intro 2,3
6.	$\left(\begin{array}{l} \mathsf{Person}(enemy) \wedge \mathsf{Saw}(enemy, killing) \wedge \\ \mathsf{Identifies}(enemy, accused) \end{array} \right)$	$\wedge$ intro 5,4
7.	$\exists z \left(\ \mathsf{Person}(z) \wedge \mathsf{Saw}(z, killing) \wedge \mathsf{Identifies}(z, accused) \ \right)$	$\exists$ intro 6,1

Fig. 7.1 Somebody says he did it

To find our man guilty, the court has to be persuaded that he fits into the framework of the **legal axiom** of Table 7.1, which I write with three different kinds of brackets so that you can more easily see its structure (don't forget that $\rightarrow$ binds right-to-left, so that $A \rightarrow B \rightarrow C$ means $A \rightarrow (B \rightarrow C)$). To find any person x guilty of crime y, we must hear another person z state that he/she saw the crime and identifies x as the perpetrator; and if that happens, we **must** find x guilty. (Another oversimplification. Keep pretending.)

It certainly looks as if the prosecution has a case. The court can reasonably deduce an instance of the existence part of the legal axiom (Fig. 7.1). Then it can go on to specialize the axiom itself, and extract a guilty judgement (Fig. 7.2). It seems that our man will be convicted this time. But he's been in more difficult corners. He's confident that his defence team will get him off.

7.1.2 He couldn't see me do it! The first part of *enemy*'s evidence was $\mathsf{Saw}(enemy, killing)$. But, says our man, it was a specially dark night: no moon, no street lights, and it was raining. He knows an expert who will testify that when it's that dark, nobody can see anything clearly at all. The expert will testify that $VeryDark \rightarrow \forall x(\forall y(\neg \mathsf{Saw}(x, y)))$, and he will bring evidence to establish $VeryDark$ on the night in question.

Suppose he does this: is there a contradiction? Our man hopes to establish $\neg\mathsf{Saw}(enemy, killing)$, and that directly contradicts *enemy*'s evidence. That may introduce doubt in the court's mind. Indeed he can do more: his evidence seems to show that $\neg\exists z(\mathsf{Saw}(z, killing))$, which would contradict any number of prosecution witnesses who might be lined up ready to testify. The reasoning is shown in Fig. 7.3. That looks like a good defence. If only the 'expert' didn't look like his twin brother ...

1.	actual *accused*	we can see him
2.	$\mathsf{Person}(accused)$	we can see that
3.	actual *killing*	previously agreed
4.	$\mathsf{Crime}(killing)$	we suppose
5.	$\left(\mathsf{Person}(accused) \to \forall y \left\{ \mathsf{Crime}(y) \to \exists z \left\langle \mathsf{Person}(z) \wedge \mathsf{Saw}(z,y) \wedge \mathsf{Identifies}(z, accused) \right\rangle \to \mathsf{Guilty}(accused, y) \right\}\right)$	$\forall$ elim (legal axiom),1
6.	$\forall y \left\{ \mathsf{Crime}(y) \to \exists z \left\langle \mathsf{Person}(z) \wedge \mathsf{Saw}(z,y) \wedge \mathsf{Identifies}(z, accused) \right\rangle \to \mathsf{Guilty}(accused, y) \right\}$	$\to$ elim 5,2
7.	$\left(\mathsf{Crime}(killing) \to \exists z \left\langle \mathsf{Person}(z) \wedge \mathsf{Saw}(z,killing) \wedge \mathsf{Identifies}(z, accused) \right\rangle \to \mathsf{Guilty}(accused, killing)\right)$	$\forall$ elim 6,3
8.	$\left(\exists z \left\langle \mathsf{Person}(z) \wedge \mathsf{Saw}(z,killing) \wedge \mathsf{Identifies}(z, accused) \right\rangle \to \mathsf{Guilty}(accused, killing)\right)$	$\to$ elim 7,4
9.	$\mathsf{Guilty}(accused, killing)$	$\to$ elim 8,(Fig. 7.1)

Fig. 7.2 The prosecution case

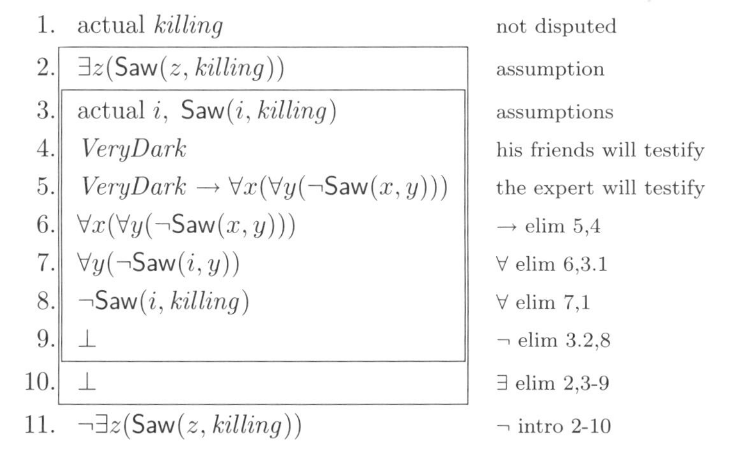

1.	actual *killing*	not disputed
2.	$\exists z(\mathsf{Saw}(z, killing))$	assumption
3.	actual i, $\mathsf{Saw}(i, killing)$	assumptions
4.	$VeryDark$	his friends will testify
5.	$VeryDark \to \forall x(\forall y(\neg\mathsf{Saw}(x,y)))$	the expert will testify
6.	$\forall x(\forall y(\neg\mathsf{Saw}(x,y)))$	$\to$ elim 5,4
7.	$\forall y(\neg\mathsf{Saw}(i,y))$	$\forall$ elim 6,3.1
8.	$\neg\mathsf{Saw}(i, killing)$	$\forall$ elim 7,1
9.	$\bot$	$\neg$ elim 3.2,8
10.	$\bot$	$\exists$ elim 2,3-9
11.	$\neg\exists z(\mathsf{Saw}(z, killing))$	$\neg$ intro 2-10

Fig. 7.3 The darkness defence

1. actual *friend*	evidently
2. $\neg\mathsf{Saw}(\mathit{friend}, \mathit{killing})$	so *friend* says
3. $\exists z(\neg\mathsf{Saw}(z, \mathit{killing}))$	$\exists$ intro 2,1

Fig. 7.4 The non-witness defence

7.1.3 My friends didn't see me do it! If the defence of darkness won't work, perhaps there's another way to contradict the damning evidence that our man was seen doing the deed. In the words of the old joke:

> *A man is accused of murder. The prosecution brings two witnesses who saw him do it. He brings five who didn't see him do it.*

There is no defence in numbers. Just one plausible witness for the prosecution will outweigh any number of non-witnesses. That's because bringing a witness who didn't see him do it — $\neg\mathsf{Saw}(\mathit{friend}, \mathit{killing})$ — isn't a direct contradiction of $\mathsf{Saw}(\mathit{enemy}, \mathit{killing})$.

Even if you turn *friend*'s evidence into an existential (Fig. 7.4) it does you no good. $\exists x(\neg P(x))$ doesn't contradict $\exists x(P(x))$. The negation's in the wrong place: you need $\neg\exists x(P(x))$, as in Fig. 7.3. A not-white swan contradicts the notion that all swans are white; it doesn't contradict the notion that some swans are white. This is a hopeless defence, and our man would be well advised not to try it.

7.1.4 It wasn't me! He did it himself! A gangster's friends respond to pressure to give helpful testimony. So long as the court doesn't recognize them and recall their past exploits, they may even be believed. Suppose a particularly helpful friend is prepared to say that he saw *enemy* doing the killing. Then surely there will be a contradiction.

Not quite: the defence first needs to persuade the court that only one person did the evil deed. It should have been careful to establish this when re-examining *enemy*. If it agrees that only one person did the killing, the court must accept an identity theorem:

$$\forall x(\forall y(\mathsf{Guilty}(x, \mathit{killing}) \wedge \mathsf{Guilty}(y, \mathit{killing}) \rightarrow x = y))$$

Given that theorem the defence can argue that sightings of two killers are in fact sightings of the same killer. You know the sort of argument: he saw the man in the black hat do it; she saw the man in the white cowboy boots do it; only one person did it; so the man in the hat and the man in the cowboy boots must be the same person.

1.	actual $accused$	visible in the dock
2.	actual $enemy$	visible in court
3.	$\mathsf{Guilty}(accused, killing)$	from $enemy$'s evidence
4.	$\mathsf{Guilty}(enemy, killing)$	from $friend$'s evidence
5.	$\forall x(\forall y(\mathsf{Guilty}(x, killing) \land \mathsf{Guilty}(y, killing) \to x = y))$	both sides agree
6.	$\forall y \left(\begin{array}{l} \mathsf{Guilty}(accused, killing) \land \mathsf{Guilty}(y, killing) \to \\ \quad accused = y \end{array} \right)$	$\forall$ elim 5,1
7.	$\mathsf{Guilty}(accused, killing) \land \mathsf{Guilty}(enemy, killing) \to accused = enemy$	$\forall$ elim 6,2
8.	$\mathsf{Guilty}(accused, killing) \land \mathsf{Guilty}(enemy, killing)$	$\land$ intro 3,4
9.	$accused = enemy$	$\to$ elim 7,8

Fig. 7.5 The third man defence

By reasoning similar to the prosecution's, *friend*'s testimony leads to the conclusion $\mathsf{Guilty}(enemy, killing)$ (gasps in court!). Then two steps of $\forall$ elim and a little juggling with connectives and we can derive $accused = enemy$ (Fig. 7.5). But the court can see $accused \neq enemy$: one is in the dock, the other is in the public gallery; contradiction! Contradictions are to be expected when gangsters fall out. Maybe that defence will work.

7.1.5 Will he get off? I hope not. He's a pretty unsavoury character, and he's got away with too much in the past. I hope he tries all three of his defences: the first and the third contradict each other, and the second is patently ludicrous. But his lawyer may be persuasive enough to make him stick to one defence, and clever enough to find one that I haven't thought of. The English court tradition doesn't seek truth, it seeks proof and often has to confront contradiction. But at least it is used to dealing with rich liars; maybe this time they will get him bang to rights.

On the other hand, perhaps he's innocent after all. Let's not be prejudiced; let's not rush to condemn. One day it might be you in the dock, and you will want to face an unbiased court.

7.2 Examples and counter-examples

The science of **cryptozoology** is about creatures that are hard to find, creatures that may or may not exist. Cryptozoologists follow trails, collect droppings, study habitats, consult local experts, but they convince the public only when they show live examples or really good photographs of 'lost' animals. The proofs of the existence of the coelecanth and the okapi, for example, were simple and utterly convincing: they showed us dead bodies and photographs and eventually, after more search, live animals. A capture of the Loch Ness monster would prove its existence; disproof, on the other hand, will require some amazing imaging equipment capable of illuminating every drop of water in the loch all at once. Even then it won't convince everybody — perhaps Nessie was on holiday in Norway on the day that we looked.

In general it is easy to persuade me to accept an existential $\exists x(P(x))$ if you can point to an **example**,[1] an individual i which has property P. Logically, that's a single-step proof with $\exists$ intro. If there are several examples, no problem: just point to any one you like. To persuade me to reject an existential, though, you have to prove its negation, using $\exists$ elim to make a schematic hypothetical proof that ends with contradiction, as for example in Fig. 7.3. That's hard: the audience tends to get lost, or lose interest, or withdraw belief before you get to the punchline. It's much easier to prove an existential than to disprove it.

With universals it's the other way round. To disprove a universal $\forall x(P(x))$ you only need to point to a single **counter-example**, an individual i which **doesn't** have property P. A step of $\forall$ elim then shows that i ought to have the property; we have a contradiction; and the public does seem to understand that argument! To prove a universal, on the other hand, needs a schematic hypothetical proof using $\forall$ intro; the public is generally unhappy about the subtlety of those proofs and in particular the level of abstract generalization. So it's much easier to disprove a universal than to prove it.

Goldbach's conjecture, for example, (every even integer $n > 2$ is the sum of two primes) would be completely, finally and convincingly disproved if a rogue integer could be found. A proof, if it ever comes, will probably be as distant from public comprehension as Wiles's proof of Fermat's last theorem.

7.2.1 Must we be agnostic about unproved claims? There have been many attempts to use logic to prove the existence or non-existence of God (the problem inspired George Boole to invent his logical algebra, for example). No-one has yet succeeded in settling the argument either way, in the sense of producing a proof which is universally, or even widely, accepted.

In this book I do not inquire into the substantive question, because I'm concerned with argument rather than reality. I'm aware that even looking at an argument may raise eyebrows, because it is a sensitive area for many people. Nevertheless I think it is possible to consider, without causing offence, one historical dispute where neither party supposed that the question was at issue, and the dispute was about the status of logical argument.

Bertrand Russell (the one who showed Frege the paradox in Chapter 1) was an atheist. He was challenged on this point. The challenger said, in short,

[1] Usually we would say **witness** but in this chapter that could be confused with witnesses at a trial.

that since Russell could not prove the **non**-existence of God, he ought to be an agnostic.

We may take it that the challenger couldn't prove the existence of God to Russell's satisfaction. We can be sure that Russell accepted that he could not prove the non-existence of God to his own or anybody else's satisfaction. In principle the challenge is "I can't prove E, you can't prove $\neg E$; you must admit that you simply don't know about E". This seems an unassailable position. Agnostic means 'not knowing': surely Russell should declare himself unsure and not pretend to be certain.

Russell was not only a logician, he was a savage wit. His reply, as short as the challenge and intentionally offensive, was that neither could he prove the non-existence of Santa Claus. To understand its logical content we must analyse Russell's reply. To see the feeling that it conveys, we must look at its implications.

First, we have to accept that it is very easy to prove the existence, at a particular point in history, of a particular live individual: for example, Richard Bornat (me). You simply point to him (me!), and everybody present will agree that he exists. You can produce him (me!) as an exhibit in a courtroom, if his existence is disputed. You can prove $\exists x(\mathsf{RichardBornat}(x))$ just by pointing at him (me!). The same goes for any of us, and the same goes for Santa Claus, **if he exists**: just bring him into the room and show him to us. That is, if it is possible to prove $\exists x(\mathsf{SantaClaus}(x))$ at all, it will be by pointing.[2] His physical presence would be a knock-down argument for his existence and a knock-down counter-example against the claim that he doesn't exist.

But how could I **disprove** the existence of somebody, like Santa Claus, who nobody has ever seen and who most of us believe doesn't exist? With great difficulty: I would have to look **everywhere** and fail to find him **anywhere**. I would have to look everywhere **all at once**, because he might flit from place to place, evading my search. But even if I discount flitting — after all, he is supposed to sleep most of the year — we haven't yet looked everywhere (under the floorboards? in the cavity wall? on the dark side of the moon? behind that cat over there? **inside** the cat?), and in practice we never could. So we can't in practice disprove the existence of Santa Claus. Putting it as a logical formula, we can't in practice prove $\neg\exists x(\mathsf{SantaClaus}(x))$ or the equivalent claim $\forall x(\neg\mathsf{SantaClaus}(x))$.

Russell's response is scientifically sound: we can't effectively disprove the existence of Santa Claus. But how is it a reply to the challenge? It implicitly admits (using 'neither') that Russell can't disprove the existence of God, and doesn't care, and then implies that belief in God is on a par with belief in Santa Claus.

[2] I wouldn't accept sleigh tracks and reindeer droppings as evidence for the existence of Santa Claus. Would you?

That insult is what Russell really intended, I'm sure. But his reply implies a question: does the challenger expect Russell to be agnostic about the existence of Santa Claus? Clearly he expects the challenger to answer "no" because, like most adults, he actively disbelieves in Santa Claus. If that is the answer Russell receives, he could then observe that the challenger accepts that it is sometimes reasonable to take a definite position on questions that cannot be decided scientifically. He might add that we all do this sort of thing all of the time, that it's a necessary point of mental hygiene not to agonize over absurd but undecidable questions.

The dispute could go no further. Fundamentally, what the challenger and Russell disagreed about was the importance of the original question. The challenger, I suspect, believed that the existence of God was too important a question to be decided on inconclusive evidence; Russell, I am certain, did not.

7.2.2 You're in the dock. "It is impossible to prove a negative", says the maxim. Actually it isn't impossible in general, but in certain cases...

Suppose that, like the gangster in the early part of this chapter, you are on trial. The prosecution brings no witnesses, but challenges you to prove your innocence.

If you can show that **nobody** could have committed the crime, by displaying an internal contradiction — perhaps the 'victim' is still alive, perhaps the evidence requires you to be in two places at once — then you can get off, as in Fig. 7.3, by showing that the assumption that somebody did the crime leads to a contradiction. You might get off if you can prove that somebody else did it, as in Fig. 7.5, if the assumption that you did it as well introduces a contradiction.

Otherwise you are in trouble. You are in very deep trouble indeed if the prosecution won't say what you accused of. You are reduced to the absurdity of the gangster's second defence: none of your friends saw you commit a crime. But then Russell's Santa Claus argument rises up against you: your friends don't watch over you 24 hours a day. You have to prove a universal, that there is **no** crime you have committed, **ever**. There could never be enough evidence to make such a proof, because it would mean examining every minute of your life so far.

This is the stuff of nightmares, of Kafka's *The Trial*. It's terrifying. It's the reason why most jurists agree that prosecutions ought not to be able to require you to prove your innocence. Unfortunately it does happen sometimes. It's happening now, in 2005. You might like to think how you could defend yourself if you were in court accused of a terrorist offence, with no right to know the nature of the offence nor any detail of the evidence against you. Perhaps the law really ought to be based on logic.

7.3 Is humankind necessarily condemned to misery?

Malthus, in *An Essay on the Principle of Population*, first published in 1798 CE, makes an argument which reaches the conclusion that most people in the world will always be starving, and there is nothing we can do about it. That's a conclusion which I and many others don't wish to believe but, like it or not, it's a powerful argument and it rumbles on to this day. I've therefore chosen to analyse Malthus's original summary of his position. All the text is original, but I've truncated it a little.

0. I think I may fairly make two postulata.
1. First, That food is necessary to the existence of man.
2. Secondly, That the passion between the sexes is necessary and will remain nearly in its present state.
3. These two laws, ever since we have had any knowledge of mankind, appear to have been fixed laws of our nature...
4. Assuming then my postulata as granted, I say, that the power of population is indefinitely greater than the power in the earth to produce subsistence for man.
5. Population, when unchecked, increases in a geometrical ratio. Subsistence increases only in an arithmetical ratio. A slight acquaintance with numbers will shew the immensity of the first power in comparison of the second.
6. By that law of our nature which makes food necessary to the life of man, the effects of these two unequal powers must be kept equal.
7. This implies a strong and constantly operating check on population from the difficulty of subsistence. This difficulty must fall somewhere and must necessarily be severely felt by a large portion of mankind.
8. Through the animal and vegetable kingdoms, nature has scattered the seeds of life abroad with the most profuse and liberal hand. She has been comparatively sparing in the room and the nourishment necessary to rear them. The germs of existence contained in this spot of earth, with ample food, and ample room to expand in, would fill millions of worlds in the course of a few thousand years. Necessity, that imperious all pervading law of nature, restrains them within the prescribed bounds. The race of plants and the race of animals shrink under this great restrictive law. And the race of man cannot, by any efforts of reason, escape from it. Among plants and animals its effects are waste of seed, sickness, and premature death. Among mankind, misery... [Misery] is an absolutely necessary consequence...
9. This natural inequality of the two powers of population and of production in the earth, and that great law of our nature which must constantly keep their effects equal, form the great difficulty that to me appears insurmountable in the way to the perfectibility of society.... No fancied equality, no agrarian regulations in their utmost extent, could remove the pressure of it even for a single century. And it appears, therefore, to be decisive against the possible existence of a society, all the members of which should live in ease, happiness, and comparative leisure; and feel no anxiety about providing the means of subsistence for themselves and families.

Malthus isn't using modern language, but he is reasoning in a way that we can recognize. He isn't calling on authority or precedent. He is trying to be scientific, and is appealing to a readership which understands 'numbers' (I would say arithmetic; school students might say algebra). He uses more premises than he seems to but, as we've already seen in the school-run example of Chapter 5, that's not unusual when rational argument is summarized.

Let's examine his argument step by step.

Para 1: food is necessary to the existence of humans. Hard to disagree with that.

Para 2: Malthus is claiming there is a desire within all people to produce children. Most people in England up to Malthus's day had indeed produced lots of children. Malthus ascribes this (as might a modern biologist) to an innate reproductive drive.

Para 3 points to evidence for the premises: history is on my side, says Malthus; life has always been like this.

Para 4 is pre-summarizing the argument to be made in the next few paragraphs.

Para 5 introduces new premises in its first two sentences; Malthus is stating them as if they were axioms. Population increases geometrically, he supposes; subsistence linearly. Population goes upwards in an geometric curve; food production goes up in a straight line. Then there is an arithmetic axiom: no matter how steeply the line is angled, no matter how shallowly the curve is angled to start with or how gently it is curved, "a slight acquaintance with numbers" (an appeal to arithmetic — your maths knowledge ought to be sufficient for you to check that he's right) will show that the population curve will grow more and more steeply, eventually more steeply than the food production line, and will get steeper and steeper for ever unless something intervenes to stop it. If it were possible, the population line would overtake the food production line, and the gap between them would widen ever faster as population growth accelerated.

So Malthus has stated **three** premises: we need food to survive; human population expands geometrically; food production can only expand linearly. He has derived from the last two, plus an axiomatic arithmetic principle about geometric and arithmetic series, the conclusion that population increase has the potential to overtake food production.

Para 6: contradiction! Population can't overtake food production, because of the first premise.

Para 7 is another pre-summary, claiming that some force is operating at the end of the 18th century CE to keep population within the bounds of food production, and that its effect is not pleasant.

Para 8 argues that the restraint proclaimed in para 7 is a force of nature, external to human society, acting just as natural forces keep animal and plant populations down. We observe, says Malthus, that animal and plant populations are kept within the limits of space and food resources by nasty means (not every seed can germinate, not every plant/pup can grow, disease or predators or simple starvation do the damage).[3] A similar force is stopping people exceeding the carrying capacity of the world, he claims, and we can observe it most clearly in the visible and widespread misery of the mass of the population.

The observed late-18th/early-19th century misery amongst the mass of the English population is not accidental, concludes Malthus. It's necessary (para 8), unavoidable and permanent (para 9). The rest of the book expands on his argument, and troubles those of us who hope that it might be possible to sustain a society in which everybody could live in 'ease, happiness, and comparative leisure'.

I'm not going to debate whether Malthus's conclusion — that most people will always live in misery — is true or false: this book is about argument, not truth. Conditions have improved beyond recognition in rural England, certainly, but to judge his conclusion nowadays we would have to consider the world as a whole, where misery is still widespread. Neither am I going to argue with his premises as stated so far, though I observe that population growth and agricultural production are still central political controversies. I'm concerned instead with his argument. Has he established his conclusion, given his premises?

I think the first part of his argument can be expressed as in Fig. 7.6. I haven't shown the reasoning which leads from line 1 to line 4, but I think the step is valid. So far so good for Malthus: human population is limited, given his premises. He has used classical contradiction to reach line 9, because he believes that either human population is limited or it isn't; quite unexceptionable in 1798, and I don't think that even now I would want to erect a constructivist objection to it.

Fig. 7.7 is a first attempt to summarize the next part of his argument. The summary, unfortunately, isn't valid. The step from box 10–13 to line 14 is the problem (never mind if you accept line 14 without argument — we aren't debating truth). You can't generalize from a particular example, or even a large number of examples, to a universal claim. You have to generalize from an abstracted example; instead, the deduction on lines 10–12 has appealed to our knowledge of animal and plant populations.

But perhaps I haven't summarized Malthus properly. He's making an argument by analogy. We can see what happens to **those** animals over there; well,

[3] When I was a child, textbooks used to contain illustrations which claimed that the offspring of a single North Sea codfish, if they all lived to maturity, would in three years or so fill the English Channel from top to bottom, side to side and end to end with no room for water. That didn't happen, of course: even in natural circumstances, only a few offspring survive each year. Now, because of overfishing, North Sea codfish are an endangered species. I expect Malthus would see that as evidence for his argument.

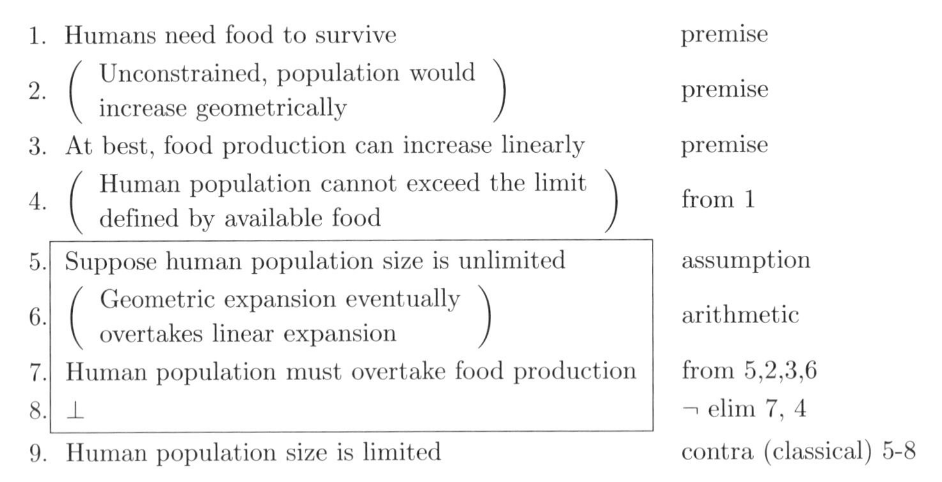

1.	Humans need food to survive	premise
2.	(Unconstrained, population would increase geometrically)	premise
3.	At best, food production can increase linearly	premise
4.	(Human population cannot exceed the limit defined by available food)	from 1
5.	Suppose human population size is unlimited	assumption
6.	(Geometric expansion eventually overtakes linear expansion)	arithmetic
7.	Human population must overtake food production	from 5,2,3,6
8.	$\perp$	$\neg$ elim 7, 4
9.	Human population size is limited	contra (classical) 5-8

Fig. 7.6 Human population size is limited, claims Malthus

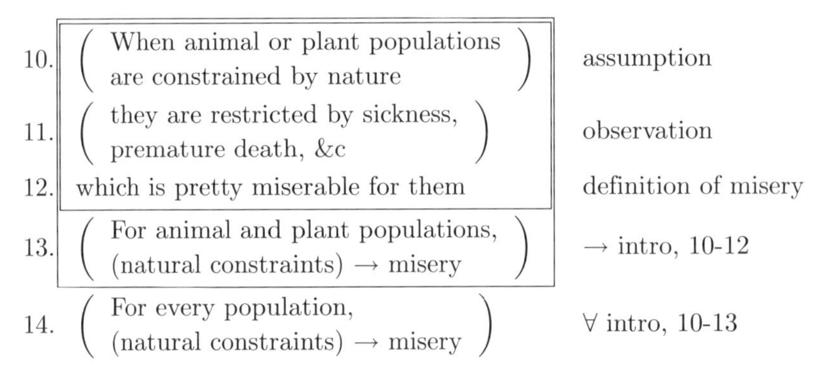

10.	(When animal or plant populations are constrained by nature)	assumption
11.	(they are restricted by sickness, premature death, &c)	observation
12.	which is pretty miserable for them	definition of misery
13.	(For animal and plant populations, (natural constraints) $\rightarrow$ misery)	$\rightarrow$ intro, 10-12
14.	(For every population, (natural constraints) $\rightarrow$ misery)	$\forall$ intro, 10-13

Fig. 7.7 A faulty attempt to show that natural constraints always cause misery

those animals aren't special: so the same thing happens to **any** animals, even us. He almost says this: "and the race of man... cannot escape from it". He's saying, perhaps, that we are animals too: a truism for a modern biologist but advanced thought for his time!

Fig. 7.8 is a second attempt to summarize the same part of his argument, this time recognizing the appeal to animality. This presentation is more acceptable, and I believe that it captures what Malthus intended. It needs yet another premise, on line 16, which we haven't seen before.

To complete his argument Malthus appears to claim that in his time, human population was under natural (resource-limited) constraint, that it was already up to the carrying capacity of the land and nothing much can be done about it ("No fancied equality, no agrarian regulations ..., could remove the pressure

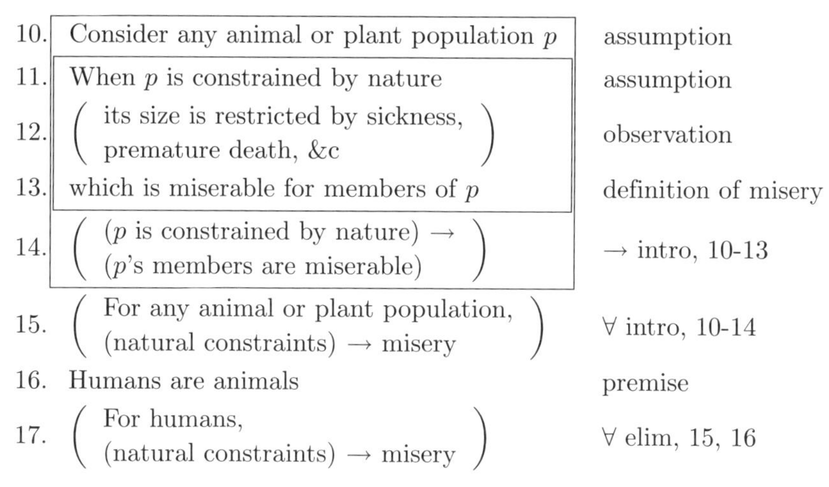

10.	Consider any animal or plant population p	assumption
11.	When p is constrained by nature	assumption
12.	(its size is restricted by sickness, premature death, &c)	observation
13.	which is miserable for members of p	definition of misery
14.	(p is constrained by nature) $\to$ (p's members are miserable)	$\to$ intro, 10-13
15.	(For any animal or plant population, (natural constraints) $\to$ misery)	$\forall$ intro, 10-14
16.	Humans are animals	premise
17.	(For humans, (natural constraints) $\to$ misery)	$\forall$ elim, 15, 16

Fig. 7.8 Natural constraints affect humans and cause misery, claims Malthus

18.	(Natural constraints are operating on human population)	premise
19.	Most humans shall be miserable	$\to$ elim, 18, 17

Fig. 7.9 Misery is inevitable, claims Malthus

of it even for a single century"). The completion of his argument, including this premise, is in Fig. 7.9.

Now, says Malthus: I've **proved** that humans must be miserable **as a consequence of natural laws**. It's easy to observe (in 1798) that they are in fact mostly pretty miserable, but I've done something different: starting from incontrovertible axioms, well-attested observations and reasoned generalizations, I've deduced that misery is unavoidable. You cannot fight the laws of numbers, Jim! Nor can you fight natural laws. They (he wouldn't say 'we', because Malthus wasn't one of the miserable many although he was by no means rich) are doomed to be miserable for ever. So don't hope for a better world, ever.

This argument is more than a historical curiosity. It informed political action. If misery is inevitable, then trying to alleviate it is pointless. Malthus's book helped inspire the oppressive Poor Laws which forced the indigent poor to work for starvation rations in workhouses, split up families and attempted to dissuade people from moving around the country to look for work. The laws themselves lasted until the 20th century, and many of the workhouse buildings still survive, taken over for use as hospitals (there's one just behind Queen Mary College in London, where I was working when I started to write this book). His argument may even have helped inspire the catastrophic work-for-food response to the Irish Famine of the mid-19th century.

We must accept, once Malthus has pointed it out, that human population would **eventually** be constrained by natural forces, with miserable results, if it continued to expand geometrically and food production couldn't keep pace. Natural constraints apply to us: we can't deny that; we are part of the natural world. But where's the evidence for Malthus's claim that the constraint was **actually operating** in 1798? Can we even be sure it is operating now?

Just like the school-run argument of Chapter 5, Malthus's argument is susceptible to attack on the basis that he hasn't proved a cause. Historically line 18 was the first point at which Malthus's argument was attacked: socialists and others argued that misery was a consequence of unfair economics, not natural constraints. It is still a point at which argument rages: some people claim, for example, that there is enough food production in the world today, that modern-day famine is caused by inadequate food distribution, and that we haven't yet reached the limits of growth.

Others attack Malthusian gloom at line 2: we need not expand population to the point of misery, they say.

The argument rumbles on. Ever since 1798 it has been a political, arithmetical, logical and sometimes even religious hot potato.

7.4 Proof search with quantifiers

Just as with connectives, formal proof with quantifiers is a lot easier than informal proof, not least because politics, religion and Santa Claus don't get a look in and I can stop trying to tell lawyer jokes.

The slogans of Table 5.2 on page 67 work well for quantifiers. In particular, you should notice that $\forall$ intro (backwards) and $\exists$ elim (forwards) should be used early, because they introduce extra assumptions, and extra individual names, into a proof. $\forall$ elim (forwards) and $\exists$ intro (backwards) use already-introduced individual names, and can safely be left till later if necessary.

In the rules of Chapter 6 I used P to stand for **any** predicate; that is, any formula with name-shaped holes in it. In examples I use R, S and T as predicate names, and I build up complex predicates using connectives and quantifiers as necessary. Just as in Chapter 5 I use E, F, G and H as simple formulae. The binding rules of Table 6.2 on page 86 show that quantifiers bind more strongly than anything.

7.4.1 Infection, subversion, transmission. Consider the claim $\forall x(R(x) \rightarrow S(x)), \forall y(S(y) \rightarrow T(y)) \vdash \forall z(R(z) \rightarrow T(z))$. Anybody who has property R

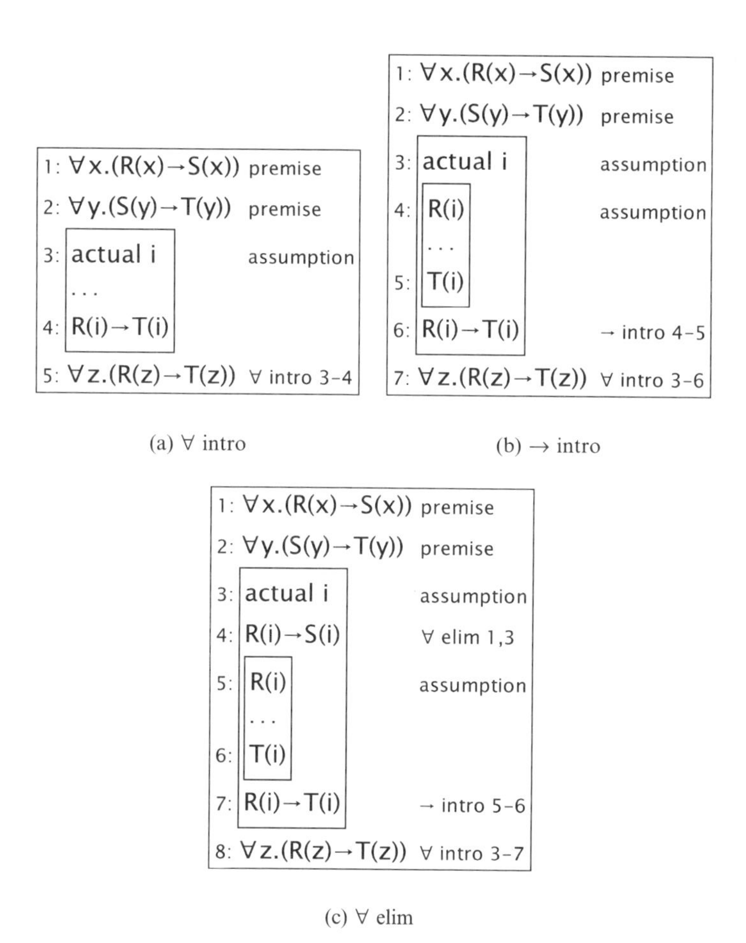

(a) ∀ intro

(b) → intro

(c) ∀ elim

Fig. 7.10 Transmission of properties

has property S; anybody who has property S has property T; therefore anybody who has property R must have property T. It surely ought to be valid, and it's a splendidly hypothetical claim: it doesn't depend on the properties of any individual; it doesn't even require that there are any individuals. It's a pure claim about transmission of properties; if you're a policeman you can also read it as a claim about the infective properties of subversive ideas.

The proof in Jape is straightforward and the first three steps are shown in Fig. 7.10. The first step, according to slogan 6 of Table 5.2 — create assumptions early — must be ∀ intro backwards. The second step, according to the same slogan, must be → intro backwards. Then we've exhausted the conclusion, and have to start working forwards. The most attractive step to take next, because it involves R, is ∀ elim using line 1 and line 3 to derive an implication. The rest of the proof is straightforward: a step of → elim extracts $S(i)$, a second step of ∀ elim extracts an implication from line 2 and line 3, and a final → elim closes.

7.4.2 Universal and existence. Consider the claim actual $j, \forall x(R(x)) \vdash \exists y (R(y))$: if there is somebody in the universe, and if everybody has property R, then there is somebody who has property R. It's pretty obvious that it ought to hold; the proof is absolutely straightforward and shown in Fig. 7.11 (I show ∀ forward, then ∃ backward, but you can do those steps in the other order if you wish). Notice that you couldn't make the proof in an empty universe because you have to appeal to the premise actual j in each of the proof steps.

The claim $\exists x(R(x)) \vdash \forall y(R(y))$, on the other hand, is absurd. The premise denies the possibility that the universe is empty (if I can point to something with property R, then there is something in the universe), but in any proof attempt the privacy conditions of the ∃ elim and ∀ intro rules make a proof impossible (Fig. 7.12): you have to reason about two different individuals which you can't make the same; $R(i)$ never proves $R(i1)$, nor vice versa.

Of course, the fact that I can't prove a claim isn't enough to show that there can't be a proof. In this case there can't be, though: in Chapter 11 I will show how to deduce from the stuck attempts in Fig. 7.12 that no proof attempt could ever succeed.

7.4.3 Green sheep in an empty field. Chapter 6 offered two arguments which showed all the sheep in an empty field could be green. In one argument everything in the field looks green —

$$\forall x(\mathsf{Green}(x)) \vdash \forall y(\mathsf{Sheep}(y) \rightarrow \mathsf{Green}(y))$$

— proved in Fig. 7.13 by using the irrelevant-implication / price-of-tomatoes trick. Note that there is no appeal to the existence of a green sheep in this proof: actual i only appears as a hypothetical assumption.

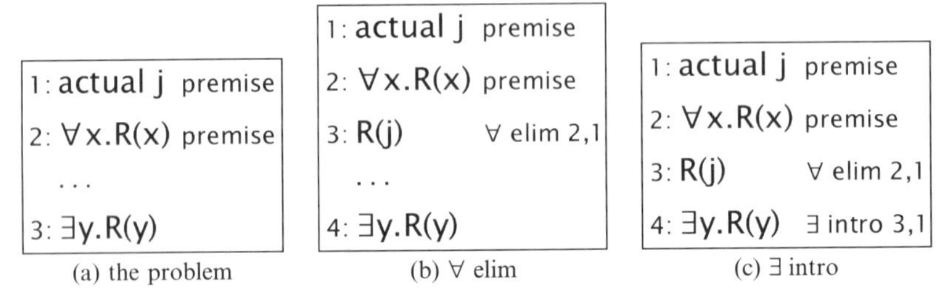

Fig. 7.11 Universal proves existence in a non-empty universe

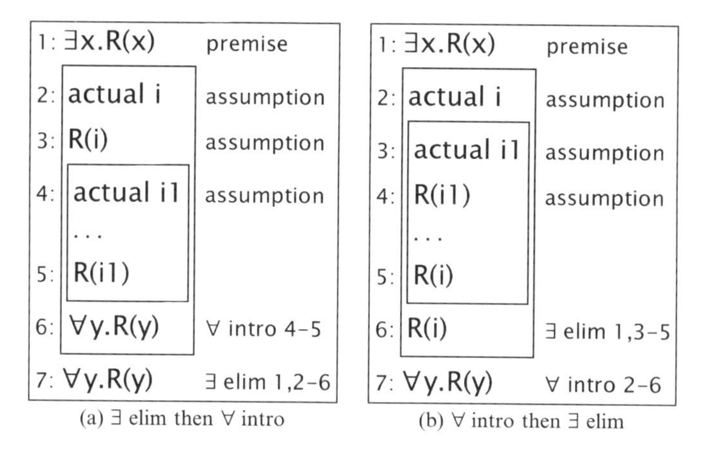

(a) ∃ elim then ∀ intro (b) ∀ intro then ∃ elim

Fig. 7.12 Existence doesn't prove universal

In the other argument the sheep can be any colour they like except green but are kept out of the puppies' pen —

$$\forall x(\neg \mathsf{Green}(x)), \forall y(\neg \mathsf{Sheep}(y)) \vdash \forall z(\mathsf{Sheep}(z) \rightarrow \mathsf{Green}(z))$$

— proved in Fig. 7.14 by using the vacuous-implication / cunning-uncle trick. This case is even worse: we require there are no sheep, but all the ones you meet are still green. (Premise $\forall x(\neg \mathsf{Green}(x))$ isn't needed in the proof, but I put it in to emphasize that the sheep outside the pen don't need to be green.)

Formally and informally, all the sheep in an empty field are green! (Also blue, purple, red, drunk, sober, microscopic, gargantuan, omniscient, ignorant, ...)

7.5 The universal drunk (a classical claim)

Consider the claim actual j, actual $k \vdash \exists x(R(x) \rightarrow R(j) \wedge R(k))$. According to the meanings of the connectives and quantifiers given in Chapters 3 and 6, it appears to say that if we pick any two individuals j and k from the universe, then we can point to an individual — not necessarily either j or k, and not necessarily neither — which, when it has property R, guarantees that j and k also have property R.

That doesn't sound controversial; indeed, it's hard at first to put it into any real-world context. But the section heading should have given you a clue. It seems to claim that there is somebody in the world who, when they are drunk, guarantees that you and I are both drunk too. Since I'm a very sober individual and I'm sure you are too, this is a very shocking claim.

Of course no such individual can exist! If you try to prove the claim constructively, not using classical contradiction anywhere, you get stuck in the sort of dead end illustrated in Fig. 7.15.

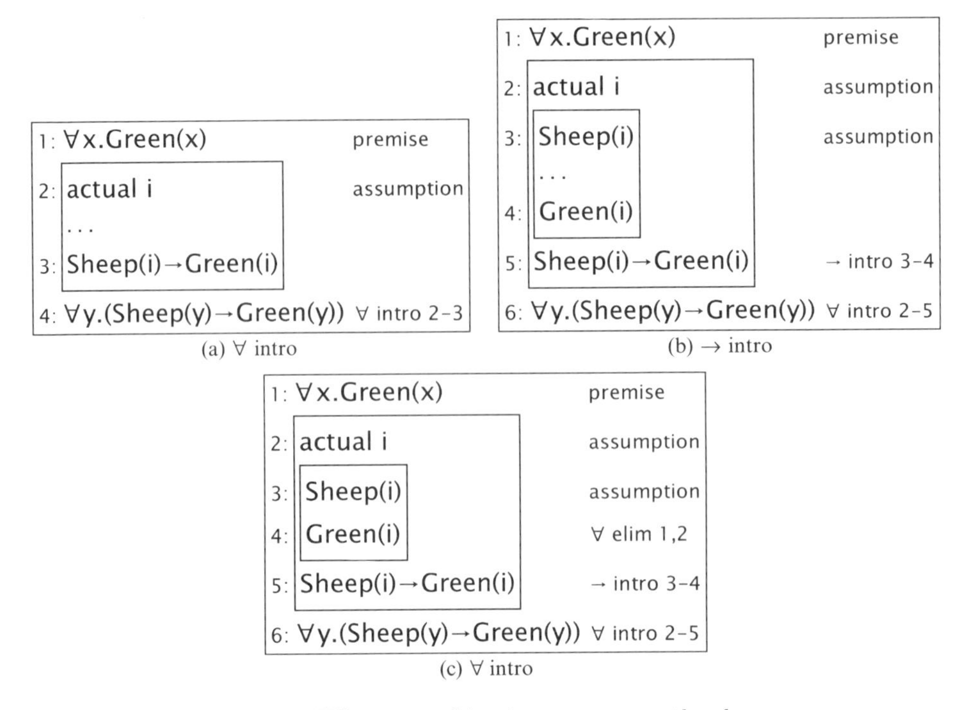

Fig. 7.13 When everything is green, so are the sheep

Disproof by dead-end isn't conclusive — just because I can't prove it doesn't mean it can't be proved — but in Part III it is shown that the universal drunk claim has a constructive disproof, and so there can't be a constructive proof. It would seem that there is no universal drunk after all.

But there **is** a classical proof — see Fig. 7.16. You start by supposing that there isn't a universal drunk, and show that this leads to a contradiction. At first the proof seem to be in the same dead end that caught the constructivists (7.16(a)) but a second contradiction step springs the trap — I used a constructive contradiction, but either kind will do — and then completion is straightforward.

Who's the drunk? Do classicists believe in a universal drunk? Certainly not! Their objection, despite the proof in Fig. 7.16, rests on the interpretation of $\exists$. The definition I gave in Chapter 6 starts "I can point to ..." in order to explain what it means to have a proof of $\exists x(P(x))$. Notice: "I can", not "I will". Classicists read it as "it is possible to point to ...", and ask us to accept $\exists x(P(x))$ when we know there must be a witness but we don't know who it is. (This is in line with the classical position that a formula is either true or false, whether we know it or not, and in opposition to the constructive position that we can only know whether we can prove it or disprove it.)

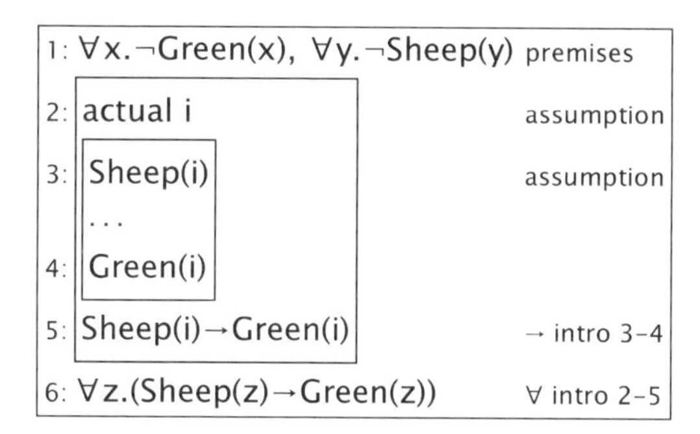

(a) ∀ intro, →intro

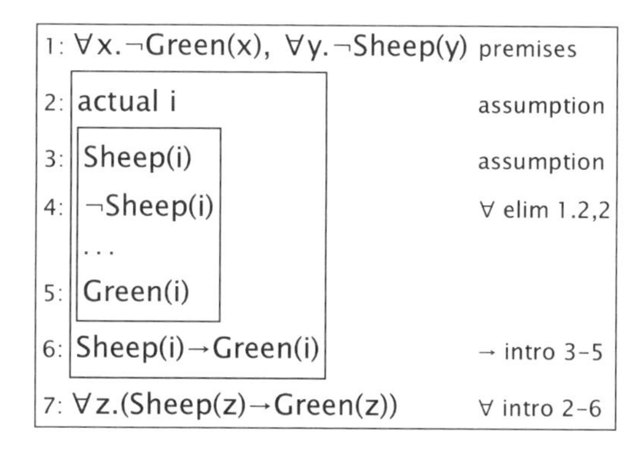

(b) ∀ elim

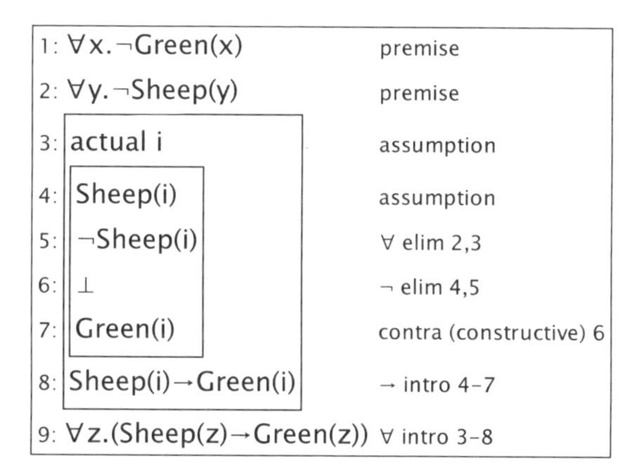

(c) ¬ elim, contradiction

Fig. 7.14 When all sheep are absent, all present sheep are green

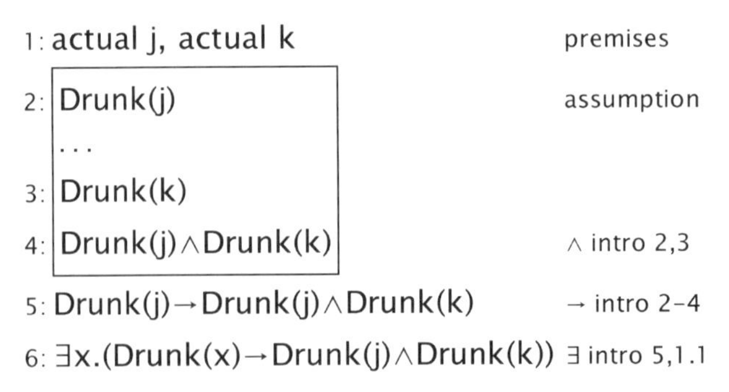

Fig. 7.15 Constructivists can't find the universal drunk

Consider, for example, a classicist might say,[4] the radio on my desk. It runs on electricity. It contains a fuse, a little bit of especially thin wire which is intended to 'blow' (melt) if too much electrical current enters the set. If it blows quickly enough then the sensitive electronics inside the radio won't be damaged. In England the plugs which go into the mains socket on the wall each have their own fuse; there is one in the distribution board which connects the sockets to the meter; there is another one still which stops too much current flowing from the electricity company's cable into the meter; and I know there are lots more in the chain that connects me, eventually, to the generation station. Some of these 'fuses' are mechanical trip-switches, but the principle is the same. Most of them are intended to stop fires in faulty cables; only the little one in the radio is there to protect sensitive and expensive electronics.

Suppose that something goes wrong somewhere in the supply chain, and somehow my electricity supply voltage doubles. Voltage is a kind of electrical pressure: press harder and more current flows. Double voltage will give me far too much current: something is going to blow. If I'm lucky it's a fuse; if I'm unlucky it's my radio. I don't know what exactly will go, says the classicist, but I'm absolutely certain that **something**'s on the way out. So I can claim

$$\exists x((\mathit{Fuse}(x) \vee \mathit{Radio}(x)) \wedge \mathit{Blow}(x))$$

— even though neither you nor I can say in advance which component it will be. Sit back and watch the fireworks!

Back to the universal drunk. Consider those two idiots j and k, next Saturday night, says the classicist. There are actually only four possibilities:

(a) they are both drunk, in which case (price of tomatoes, choose a tomato) $\mathit{Drunk}(j) \rightarrow \mathit{Drunk}(j) \wedge \mathit{Drunk}(k)$ or $\mathit{Drunk}(k) \rightarrow \mathit{Drunk}(j) \wedge \mathit{Drunk}(k)$;

(b) j is sober but k is drunk, in which case (cunning uncle) $\mathit{Drunk}(j) \rightarrow \mathit{Drunk}(j) \wedge \mathit{Drunk}(k)$;

[4] Thanks to Thomas Forster for this example.

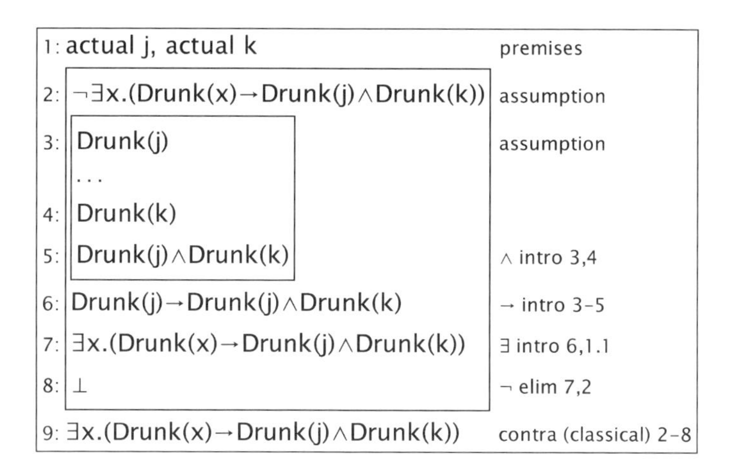

(a) not a dead end

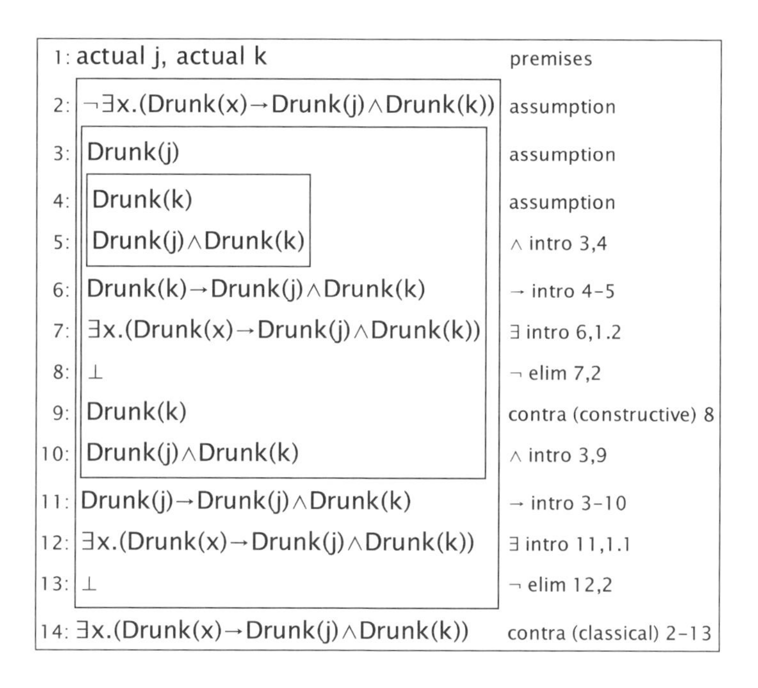

(b) the proof complete

Fig. 7.16 A classical universal drunk

(c) k is sober and j is drunk, in which case (the other uncle) $Drunk(k) \to Drunk(j) \wedge Drunk(k)$;

(d) they are both sober, in which case (pick an uncle) $Drunk(j) \to Drunk(j) \wedge Drunk(k)$ or $Drunk(k) \to R(j) \wedge R(k)$.

So **no matter what they do** there will be somebody to blame, somebody to put on the left of the implication. For a classicist, the claim isn't that there is a particular universal drunk, rather that something always turns up. It's a consequence of the classical treatment of implications and existentials, which is discussed in more detail in Chapter 10.

An eternal abstainer? Of course there can't be a universal drunk. There might, however, be a person who is never drunk — $\exists x(\neg Drunk(x))$. We could then say (cunning aunties and uncles that we are) that if that saintly person is ever drunk, so are the rest of us. If that were added as a premise, then constructivists would have no difficulty proving the conclusion. That version of the claim is in Jape's Conjectures panel, so you can try it for yourself.

So what? Luckily for us, logic isn't nonsense. Even classical logic doesn't demand that we believe in a universal drunk. But there is a lesson to be learnt; it's more than an academic dispute.

If logic is about proof, then you can claim less — support fewer claims — than you can if it is about truth. Programmers, it seems to me, have to rest their claims on proof. When we produce a piece of software we need to know that it works, and so far all the ways that we can do that require us to show **why** it works — that is, provide a proof. If we can't prove it, all we have is statistics based on testing. Any user of commercial software in the first decade of the 21st century knows how awful a prospect that is.

Proof matters, which is why so much of this book is about proof.

Part III

Disproof

Formal logical proof is a mathematical puzzle game: "here is a logical claim; there are some formal rules; can you prove the claim using the rules?". The game can be fun, and it's the game you have to play when you are programming a computer. But if that's all there was to logic, it wouldn't be much.

Logic is more than a game because logical proofs mirror practical, useful reasoning. It says something about the way we reason **and** it's a game which machines can play. It's only useful to make machines play the game because of its connection to valid reasoning.

We can challenge a logical claim in two different ways. "Can you prove it?" is the puzzle game of Part II. "Is that so?" is the real world view explained in this part. If logic is to mirror our world then every claim that's logically provable must be real-world undeniable, and vice-versa. If there are gaps or disagreements then either the logic is not as logical as it should be or we don't understand the world as well as we think we do.

You can deny a logical claim if you are in a **situation** in which the claim obviously doesn't hold. The claim "it's Tuesday", for example, is deniable six days every week. More abstractly, the claim E is deniable in a situation where there can be no proof of E. We don't have to think of concrete Es; we don't have to live in such a situation, only be able to imagine it without generating a contradiction. That particular imaginary situation, or **possible world**, is a **counter-example** to the absurd claim that E holds always and everywhere, and therefore ought to be provable. On the contrary, we can show that it doesn't always hold, so it ought not to be provable!

The method of counter-examples turns out to be the easiest and most convincing way to **disprove** an invalid logical claim. Consequently, this part of the book is about disproof by counter-example. Proof by logical argument; disproof by counter-example. That's how it's done.

Our counter-examples, to be utterly convincing, have to be precise and unanswerable; for our purposes that means they must be mathematical. Chapter 8 explains how the mathematics of disproof fits with the mathematics of proof. Chapters 9 and 10 describe the kind of arguments which count as mathematical descriptions of situations in our two kinds of logic — constructive and classical, respectively — and thus serve as examples of or counter-examples to logical claims. Chapter 11 shows various ways in which you can generate disproofs of unprovable claims.

8 Disproof in a mathematical model

Fig. 8.1 shows a Jape proof attempt which is in trouble. A raw novice, afraid to use a backwards step from line 4, has made an $\rightarrow$ elim step from $E \rightarrow F$ on line 1 to F on line 3.[1] Jape has loyally made the step, but the $\rightarrow$ elim rule demands a proof of E on line 2. If you ever wander into this logical blind alley you will find no way out but Undo or Quit.

The gap between lines 1 and 2 asks the impossible: nobody has ever shown and nobody will ever show, using the rules of Natural Deduction, that E is a logical consequence of $E \rightarrow F$. It isn't a matter of debate; there isn't a possibility that anybody might invent an ingenious solution tomorrow. Great mathematicians have shown us that it's **impossible**, in a sense that will be made clear below.

It would be nice to be able to distinguish dead ends and their Great Mathematical Obstacles from paths which are just rather more difficult than we expected. Searching for counter-examples can help: if we know we're stuck we can stop banging our heads on a brick wall, back out and look for another way through.

8.1 Counter-examples

Chapter 3 says that $E \rightarrow F$ means "whenever you accept E, you are forced to accept F". To disprove the claim $E \rightarrow F \vdash E$, which is what lines 1 and 2 of Fig. 8.1 amount to, we must show situations in which any reasonable person would accept $E \rightarrow F$, but need not at the same time accept E.

8.1.1 Informal counter-examples. One way to disprove the $E \rightarrow F \vdash E$ claim is to think of real-world Es and Fs for which the claim doesn't hold, as for example in Table 8.1. Each of the situations in the table shows that I need not always accept that every E is a consequence of $E \rightarrow F$. Since it doesn't hold in some cases, it doesn't hold **in general**. I'd be a fool to look for a proof that $E \rightarrow F \vdash E$, and a bigger fool to believe anybody who said they'd found one.

[1] Half-forward, half-backward steps like the one in Fig. 8.1 can confuse the novice. They're made difficult in Jape, because they can be confusing, but they're allowed, because they are sometimes convenient and occasionally essential.

1: E→F premise

. . .

2: E

3: F → elim 1,2

. . .

4: (F→G)→(E→G)

Fig. 8.1 A stuck proof

Table 8.1 Informal counter-examples to $E \to F \vdash E$

1. Suppose I accept that when it rains, I get wet. I need not **therefore** accept that it is raining now (even if I'm soaking wet).
2. Suppose I accept that if it is Thursday, then this must be Paris. I need not **therefore** accept that it is Thursday now (wherever I am).
3. Suppose I accept that if I ever met a grizzly bear I should be eaten. I need not **therefore** accept that you're a bear (whether or not I'm being nibbled).

8.1.2 Mathematical counter-examples. Counter-examples in English aren't quite convincing. Abstract counter-examples which don't mention rain, or Paris, or bears would be better. Better still if we can use diagrams.

We might be able to do it it with a diagram which mentions only E and F, because there are no other identifiers in the $E \to F \vdash E$ claim. Fig. 8.2 shows all the **possible worlds** which contain E, or F, or both, or neither. If even a single one of these worlds is a counter-example, the logical claim surely ought to be denied.

- In 8.2(a) we have both E and F, so we can hardly deny $E \to F$. And we have E, so we have both premise and conclusion. The claim holds in this world: i.e. it's an example.
- In 8.2(b) we can deny $E \to F$ because we have E but not F. The world is irrelevant to the claim; it's neither example nor counter-example.
- In 8.2(c), by the price-of-tomatoes argument, it is the case that "whenever you accept E, you are forced to accept F", i.e. $E \to F$. It would certainly be the case that if there were a proof of E you would accept F, because you already have a proof of F. But it's certainly **not** the case that you must accept E in that world: it's a counter-example.
- Figure 8.2(d) is another counter-example, this time by the cunning-uncle argument. You will never come across a proof of E, so the $E \to F$ promise will never be broken: i.e. $E \to F$. And, because E isn't shown, the conclusion doesn't hold: we aren't forced to accept E.

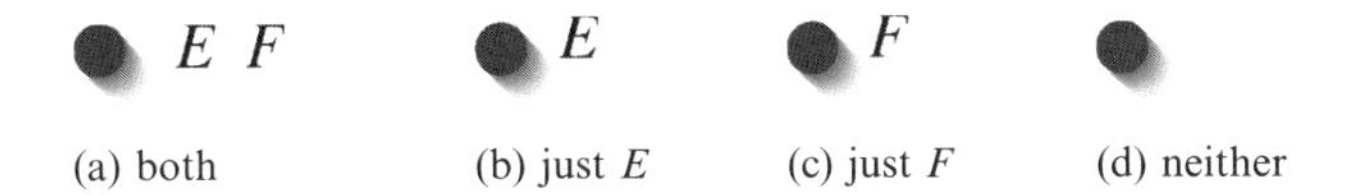

Fig. 8.2 Possible E, F worlds

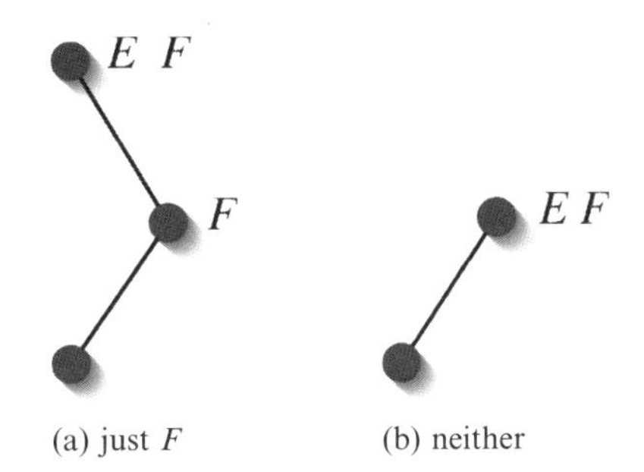

Fig. 8.3 Some possible E, F universes

So there are at least two imaginable situations in which you can't deny $E \rightarrow F$, but you can deny E. It's not the case, by demonstration, that one always leads to the other; we've disproved the claim by counter-example. But the demonstration isn't very satisfying: the counter-examples each rest on the oddities of implication, not its straightforward meaning. The nonsense of $E \rightarrow F \vdash E$ was more convincingly exposed by the sentences in Table 8.1.

If we imagine a universe of possible worlds, with transport between them, then we can construct more convincing abstract disproofs. Fig. 8.3 shows two such. In each universe it's very clear that "whenever you accept E, you are forced to accept F" because the worlds that contain E also contain F, but each universe also contains a world which doesn't mention E. These diagrams show that in a **universe** in which we are forced to accept $E \rightarrow F$, it's not the case that every **world** forces us to accept E. Figure 8.3(a) can be read like the rain example from Table 8.1: I'm wet now (middle right) and it is true that when it rains I get wet (top left) but it's not raining now (perhaps I just got out of the bath). Fig. 8.3(b) could be the bear example: I never meet any bears (bottom left); I nervously expect (top right) that if I met one I'd be eaten; but I'm not in danger right now.

Chapter 9 explains how to write and read multi-world diagrams like those of Fig. 8.3, and how to use them as counter-examples. I find them more convincing counter-examples to $E \rightarrow F \vdash E$ than the single worlds of Figs. 8.2(c) and 8.2(d) but — beware! — they are **constructive**, not classical, counter-examples. Classicists use only single-world diagrams; by considering multi-world universes, constructivists can disprove more claims. The difference is precisely the dual of the fact that classical contradiction allows more proofs than the constructive version: the extra claims you can prove are just the ones that multi-world diagrams can disprove. As you will see!

8.2 Mathematical models

Reality is what we live in. It's what's out there, what is, what happens: the stubbed toe, the rain in your face, the view from your window. Logic is a kind of mathematics, and making it correspond to lived reality isn't a trivial matter. That's why Aristotle and his Ancient Greek colleagues are so rightly celebrated: they built a bit of mathematics which corresponded to sound reasoning. Philosophers and logicians have been refining and polishing their efforts ever since.

But 'reality' is at least as slippery a notion as 'truth'. Philosophers and scientists have endless fun with it. The best we can expect is a bit of mathematics — called a **model** — that describes a view of reality, plus an assurance that **if** reality is like the model, **then** the logic won't lead you astray. For classical Natural Deduction we imagine that formulae are 'really' either true or false, the mathematical model is discussed in Chapter 10, and the logic is what was discussed in Part II, using the classical contradiction rule. For constructive Natural Deduction we imagine that proofs are 'real', the mathematical model is in Chapter 9, and the logic uses constructive contradiction.

8.3 Syntactic and semantic claims

In the illustrations above I made fast and loose with notation. I said that I'd disproved the sequent $E \to F \vdash E$. Indeed I had, but only indirectly. What the counter-examples, either words or diagrams, actually disprove is $E \to F \models E$, the corresponding **semantic** sequent. A syntactic sequent using $\vdash$ makes a claim that there is a proof; its semantic counterpart using $\models$ claims that there are no counter-examples. The turnstiles even sound different: $\vdash$ is pronounced "proves", and $\models$ is pronounced "models".

- The syntactic sequent $A_1, A_2, \ldots, A_n \vdash B$ claims that there is a formal proof which connects premise formulae $A_1, A_2, \ldots, A_n$ to conclusion formulae B. It's a formal claim to go with a formal proof.
- The semantic sequent $A_1, A_2, \ldots, A_n \models B$ claims that in **every** situation in which I'm forced to accept **all** of the premise claims $A_1, A_2, \ldots, A_n$, I will **also** be forced to accept the conclusion claim B. Just what a 'situation' is, and just what 'forced to accept' means, varies between mathematical models.

Syntactic and semantic sequents which are identical apart from their turnstiles are clearly related, although they claim different things. If model and logic are related as they should be, the two sequents make the same claim in different ways. From now on, I'll just say "claim" and let the context distinguish.

8.4 Situations as examples and counter-examples

A positive demonstration — a situation in which I am forced to accept premises $A_1, A_2, \ldots, A_n$ and at the same time forced to accept conclusion B — is an **example** of a claim, but an example doesn't amount to proof. A negative demonstration — a situation in which I'm not forced to accept conclusion B even though I am forced to accept premises $A_1, A_2, \ldots, A_n$ — is a **counter-example**, a demonstration that the connection isn't universally valid in the model, and therefore a **disproof**.

Examples can be illuminating, and some claims can even be proved by exhaustively listing possible situations. I am confident as I write this, for example, that every person listed under the name Bornat in any telephone directory in the UK is a direct descendant of my father. By restricting the claim to a particular finite collection of data I've made sure that there is only a finite collection of situations, and by pointing to printed directories I've made sure that the facts won't change while you examine them. But most interesting claims refer to more situations than you would have time to consider, sometimes even infinite numbers of situations, and then Russell's Santa Claus analogy explains why it would be futile to try to prove them by exhaustion.

On the other hand, any number — even an infinite number — of examples can be overthrown by a single counter-example. So far as this book is concerned, the purpose of a mathematical model is to allow us to demonstrate disproofs — to show, for example, that the proof attempt of Fig. 8.1 is not temporarily stuck, it's permanently done for.

8.5 Soundness and completeness

We can prove claims in a formal logic using rules; we can disprove claims in a mathematical model by demonstrating counter-example situations. What's the connection?

Suppose we have a formal logic and a mathematical model. A claim is **valid** if it's impossible to generate a counter-example situation in the model. A claim is **provable** if we can show a proof in the logic. The logic is **sound** if every provable claim is valid; it's **complete** if every valid claim is provable. Soundness means you can't prove anything you ought not to be able to; completeness means you can prove everything you ought to be able to.

Suppose you used a logic in which there were no axioms and no rules of inference at all. There would then be no proofs; in effect the logic would always

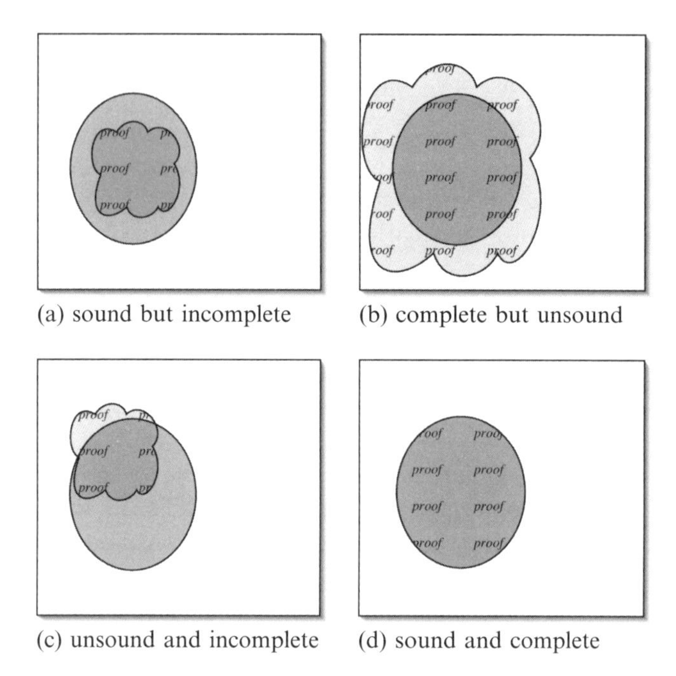

(a) sound but incomplete (b) complete but unsound

(c) unsound and incomplete (d) sound and complete

Fig. 8.4 Soundness, unsoundness, completeness and incompleteness

say "no!". It would be a sound logic, too, because it would never allow any proofs at all, and therefore there could never be a proof of an invalid claim. But, of course, it would be useless.

On the other hand, suppose you used a logic with only one axiom/rule: no matter what the claim, accept it without ado. Every claim would be provable; in effect the logic would always say "yes!". It would be a complete logic, because every claim would have a proof, and therefore every valid claim would have a proof. And it, too, would be useless.

The "no!" logic is useless because it misses all the proofs; the "yes!" logic is useless because it doesn't notice any of the disproofs. Neither system is to be trusted.

Clearly we should generate more proofs than the "no!" logic and fewer than the "yes!" logic. The ideal is that our logic should be sound **and** complete: then it will says "yes!" whenever it should (because it is complete) and "no!" whenever it should (because it is sound). A sound and complete logic will correspond exactly to its model. Natural Deduction is such a logic, when we give it the models illustrated in this chapter and explained in detail in Chapters 9 and 10.

In Fig. 8.4 I depict the universe of semantic claims and their relation to proofs. The dark oval represents valid claims and the white square all claims; so the area of the white square outside the dark oval represents invalid claims. If all provable claims, shown by a cloud with lots of 'proof' words inside it, fall inside the dark oval then the logic is sound; if all the dark oval falls inside the proof cloud then the logic is complete. Only when the proof cloud and the validity oval coincide, as in Fig. 8.4(d), does the logic have both properties at once.

The way that the sound/complete distinction is phrased treats the model as primary, and the logic as secondary. That is, indeed, the way Platonists see it: eternal truth was surely around long before we began to try to trap it in logic. On the other hand, constructive logic was invented before Kripke gave it a mathematical model, so perhaps his model should be judged rather than the logic. But in the end constructivists and classicists want exactly the same thing — a formal logic and a mathematical model that exactly correspond. Soundness and completeness describe the perfect fit.

8.6 Natural Deduction is sound and complete

Constructive and classical Natural Deduction each have a mathematical model. Each is sound and complete relative to its model. In each logic there is a proof of a claim if and only if there are no counter-examples, and conversely there is a counter-example to a claim if and only if there is no proof. The models are discussed in chapters 9 and 10. Proofs of soundness and completeness are not discussed (we leave that up to the mathematicians).

8.7 Does the exception prove the rule?

— *You are always late.*

— *Not so! I was early yesterday.*

— *Ah, but the exception proves the rule!*

"Counter-examples destroy claims" runs the argument in this chapter. But "the exception proves the rule" seems to contradict that. It sounds as if a counter-example **strengthens** a claim, clinches it, rather than undermining it.

Actually that's a misunderstanding. The proverb refers to an ancient legal principle about tacit (hidden, silent, implicit) agreements which has been known and used as far back as the Roman Empire. The principle is that evidence of special permission for (or prohibition of) an activity is also evidence for an implicit general rule to the contrary. If I give you permission to pick an apple from my tree then I tacitly claim, and by asking permission you implicitly acknowledge, that I have the right to refuse you. If my best argument is that I wasn't late yesterday, I implicitly acknowledge that in general I do arrive late. If there's a notice on only one of a university's lawns saying 'keep off the grass' then implicitly I am allowed to walk on the others.

An exception has to be a special case that breaks a rule. It's the acknowledgement that it's an exception that proves the rule exists. Counter-examples don't strengthen arguments after all. Logic is safe.

8.8 No smoke without fire?

> *The lady protests too much, methinks.*
>
> Gertrude to Hamlet, of the Player Queen.

> *Why a hundred? One would have been enough.*
>
> Albert Einstein, of the pamphlet *100 authors against Einstein.*

In life we are suspicious of too much disproof. It smells of bluster, cover-up, a dishonest guilty mind. We'd prefer an unemotional denial and let that be the end of it.

In logic, emotion isn't involved and we neither count nor weigh arguments. We don't count examples at all, and as we shall see in the next chapter there can be provable claims with no witness examples in the model. All we care about is counter-examples, and the number doesn't matter. Show one, a hundred, an infinity — it's all the same. A valid claim will have no counter-examples and an invalid claim will have at least one. Beyond that, who's counting?

9 Constructive semantics

This chapter introduces the mathematical model which lies behind constructive Natural Deduction. It's built around a simple and beautiful reading of negation and implication. Best of all, it uses diagrams.

The model itself is due to Saul Kripke, and it relates formal proof to real-world reasoning in very much the way that programming relates to reality: there are things we can do (claims we can prove), things we can't do (claims we can disprove) and things we're still trying to do (claims we can only speculate about). Inventing the model was brilliant work; the mathematics which showed that the logic is sound and complete was a work of genius, and Kripke was awarded the Fields Medal for it. This book doesn't investigate Kripke's proofs, but we can exploit his results.

Kripke's work didn't end with his model of constructive proof. He's now a very distinguished philosopher, extending his ideas into theories of semantics more generally.

9.1 Proofs are central

In this book the semantic sequent $A_1, A_2, \ldots, A_n \models B$ is read as a claim that in any situation in which I'm forced to accept hypotheses $A_1, A_2, \ldots, A_n$, I'm also forced to accept conclusion B. It's up to the mathematical model to define what a 'situation' is and what 'forced to accept' means. When we investigate claims we imagine situations in which we are forced to accept the premises and see if we are then forced to accept the conclusion as well.

From the constructive point of view, we are forced to accept those things of which we have proof, and that is the starting point of the model. We can then be forced to accept composites of which we don't necessarily have proof — for example, we're forced to accept $E \wedge F$ if we have separate proofs of E and F. The model defines how to treat each of the connectives and quantifiers, building up from atomic formulae.

In the model there are three different possibilities for a particular formula in a particular situation: there may be a proof; there may be no proof; it may be possible to show that there could never be a proof. The same extends to composite formulae: we may be forced to accept; we may not be forced to accept; we may be able to say that we could never be forced to accept.

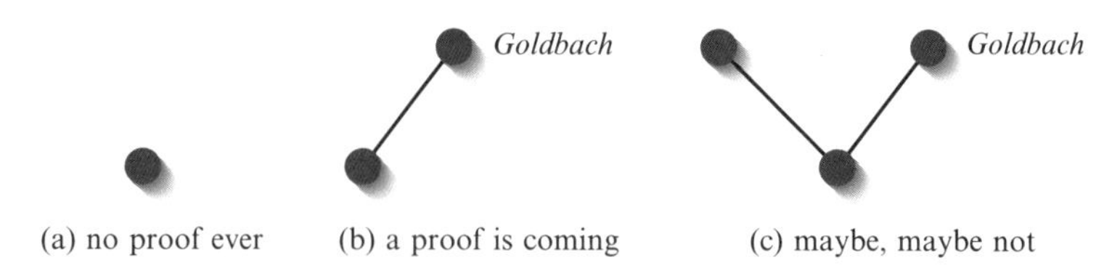

Fig. 9.1 Possible developments with Goldbach's conjecture

> Don't suppose that you could use different values — 1, $\frac{1}{2}$ and 0, say — to recognize these different possibilities. That wouldn't capture the subtleties of constructive proof at all. This isn't a three-valued logic.

The model captures the distinction between "no proof yet", which is how novel claims begin, and "no proof ever", which is the fate of invalid claims. Because constructive Natural Deduction is sound and complete in this model, every claim is either provable or disprovable — no middle ground.

Fig. 9.1 shows an example of the subtlety which multi-world diagrams give to constructive semantics. There are many ways of depicting the fact that Goldbach's conjecture — "every even integer $n > 2$ is the sum of two primes" — currently has no proof. Figure 9.1(a) illustrates the gloomy possibility that there might never be a proof, either because nobody ever finds one, or because somebody finds a disproof. Figure 9.1(b) shows what everybody hopes is the case: we don't have a proof now, but one day we will. Figure 9.1(c) hedges its bets, setting out alternative possibilities of failure and success.

The three different situations have different content, as you'd expect. In none of them are we forced to accept *Goldbach* (definitions 9.1, 9.2); only in 9.1(a) are we forced to accept $\neg$*Goldbach* (Definitions 9.5, 9.6); in 9.1(b) and 9.1(c) we have neither *Goldbach* nor $\neg$*Goldbach*, and these are each constructive counter-examples to *Goldbach* $\vee$ $\neg$*Goldbach* (Definition 9.4).

9.1.1 Positive and negative evidence. Consider $F \rightarrow E$. This isn't a claim which is valid in general: there are lots of Fs and Es I can think of which aren't connected in that way. Rainfall $\rightarrow$ unhappiness, for example, isn't always so. It would hold on a camping holiday in a leaky tent; it wouldn't hold if you were farming a field in a drought. Clearly, $F \rightarrow E$ is acceptable in some situations and not in others. In the model, if E always turns up when F does, then you certainly ought to accept $F \rightarrow E$; if F sometimes turns up on its own, you certainly shouldn't. Figure 8.3(b) on page 125 shows a situation in which it's right to accept $F \rightarrow E$, and Fig. 8.3(a) shows another in which it would be wrong. As in real life, it depends on the situation.

A model of Natural Deduction has to capture all its peculiarities. Recall $F \vdash E \to F$ (the price-of-tomatoes argument) and $\neg E \vdash E \to F$ (the cunning-uncle argument). Those two oddities influence the whole model. In particular you are forced to accept $A \to B$ unless you can find evidence to the contrary, which the model defines to be a world in which you are forced to accept A but not forced to accept B. This is called a **negative** definition — forcing in the absence of contrary evidence. Negation ($\neg$) and universal quantification ($\forall$) have negative definitions too.

> Don't suppose that the peculiarities of implication are just to do with constructive semantics — they're shared with classical logic and lots of other logics. Devising a useful logic which is useful, sound and complete and doesn't have those peculiarities is a challenge. Both the versions of Natural Deduction in this book have difficulties with implication (but the constructive version does rub your nose in it).

Negative definitions may seem peculiar at first, but they give the model the odour of disproof, which is just what we need. We are as often seeking evidence that we are **not** forced to accept some formula as the other way round. Negative definitions often suit our purposes.

9.1.2 Diehards beware! Because Kripke's model allows us to describe and investigate situations in which we may not immediately know everything there is to know, it does more than classical logic needs. It covers classical concerns in the sense that it includes the classical model: the discussion on page 116 around Fig. 8.2 and the $E \to F \models E$ claim, for example, is just like a classical treatment using truth tables (see Chapter 10). But in talking about change and the development of knowledge, the constructive model goes further than a Platonist would wish, and it allows disproofs of many Platonist claims, including $E \vee \neg E$.

If you are used to classical logic, perhaps as applied to computer hardware design via Boolean algebra, Kripke's treatment may seem perplexing at first. Hold on to your hat and keep your hands inside the car: it's an exciting ride and you'll learn a lot.

9.2 What counts as a situation?

We begin with the intuition that proofs have to be discovered, so that a proof collection builds up over time. A situation is a possible world — that is, a collection of proofs — plus all the worlds which you decide to allow it to develop

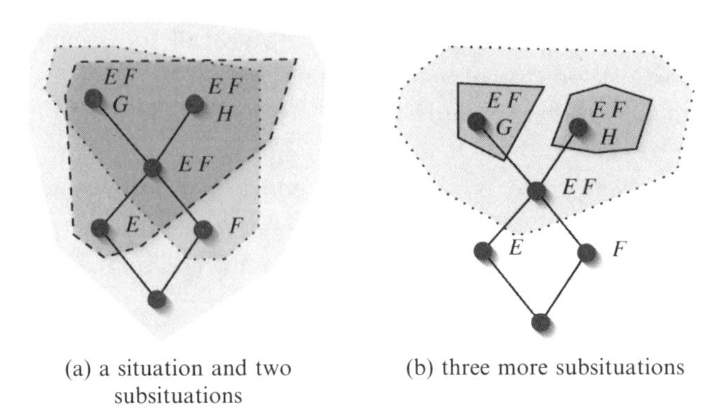

(a) a situation and two subsituations

(b) three more subsituations

Fig. 9.2 A situation and its subsituations

into. You need not allow all development possibilities: by making particular choices you present a particular challenge to a claim which ought to hold in every situation. "What", you say, "if the world was like this and developed like so?". If the claim is universally valid then it will withstand any challenge you invent. On the other hand, if you can imagine a situation where the claim falls down, then it's not universal after all — which means, so far as logic is concerned, that it's an invalid claim.

We show our imagined situations using diagrams. In principle a diagram is just a collection of **possible worlds** — blobs with associated collections of proofs — and lines between them, but there are restrictions.

9.2.1 Restricted connections. The lines in a situation are directional: they point **from** one world **to** another. They are intended to show possible development of proofs: you can draw a line from w to w' only if w' has the same proofs as w, or more proofs than w. This is the **monotonicity** condition. (There's no point having a line back from w' to w, because that would mean that the two worlds would have to have the same proofs, and then you could amalgamate them.)

Since we only need connections in one direction, we can avoid drawing arrows on the lines; instead we arrange our diagrams so that all connections point **upwards**. Each world in a diagram is then the **root** of a situation which consists of that world plus all the other worlds you can reach from it by making one or more upward moves.

In a diagram there are clearly as many situations as there are worlds. Situations may overlap, as illustrated in Fig. 9.2, if a world has more than one parent.[1]

[1] In general our diagrams are DAGs — Directed Acyclic Graphs. But in most cases worlds have single parents, situations don't overlap, and then they are merely Rooted Directed Trees.

9.2.2 Restricted claims. The formulae we write next to a world are those which we imagine have proofs in that world. We restrict the kind of formulae that we allow — no $\top$, no $\bot$, no connectives, no quantifiers, just simple identifiers like E and F, individual instances of named predicates like $R(i)$ and $S(j)$, named relations like $T(i, j, k)$ and $\mathsf{Deserves}(richard, favour)$, and presence markers like actual i and actual j. Calculations about composite claims — formulae which include connectives or quantifiers — happen off-diagram. Unfortunately this means that a lot of the action has to happen in your head, on paper or on a blackboard. We'll see how to deal with that after we've got past the basics (but don't fret: Jape can help).

9.2.3 This is what counts. There are just four conditions, summarized in Table 9.1.

Table 9.1 Requirements for a situation diagram

rooted one root world, at the bottom;
atomicity no composite claims, no $\top$, no $\bot$;
direction only move upwards;
monotonicity child worlds must include all their parents' claims.

9.2.4 Is monotonicity believable? The monotonicity principle in the constructive model — proofs are never withdrawn, never contradicted, never go out of date — is the first point at which you might raise an objection. In the real world proofs are contested, revised, withdrawn and can be refuted. In *Proofs and Refutations*, for example, Lakatos famously described the history of a conjecture of Euler's about solid objects. What counted as a proof in one century wasn't accepted in the next. Proofs had to evolve to keep up with changing notions of what counts as a solid object, as the conjecture was tested against concave solids and solids with worm-holes.

The underpinnings of a real-world proof aren't fixed either: real-world logics are disputed, modified and updated. Computer programming languages in particular are modified and re-modified over and over. When a new version of Caml or Java comes out, for example, I have to tinker with the innards of Jape to keep it running. But in the constructive model, proofs are forever: how can this be?

There isn't really a disagreement. The multi-world diagrams of this chapter describe how we can reasonably interpret particular fixed claims in a particular fixed logic. Claims like that aren't contested even in our world, provided they obey the logic's rules. Instead, people dispute that a particular problem formulation captures the matter to be proved, or they argue that a logic isn't adequate to distinguish the nuances of meaning that it should, or they don't use a formal logic at all. So "new proofs" are really of new claims and/or in new logics and/or not formal at all. I may call the various versions of Jape by the same name; strictly they are different programs in different logics.

The monotonicity condition makes sense once you realize that we are only talking about what it is like to deal with proofs of particular fixed formula shapes in a particular fixed logic, and that we aren't pretending to describe the meta-logical development of conjectures and reasoning in the real world.

9.2.5 Forced to accept *now*? In the real world we live in a continuous present. It's always "now"; we can know what happened in the past, more or less, but all we know about the future is that we don't know. We don't know what's going to happen in the next instant, let alone next week or next year. It's one thing to say that I'm forced to accept $E \wedge F$ if I have proofs of E and F to hand, and not otherwise; it's quite another thing to say that in some possible world, as for example in Fig. 8.3(b) on page 125, I can be forced to accept $E \rightarrow F$ if in future worlds I will always find F whenever I find E, even though currently I have a proof of neither. That feels as if facts about the future are acting on knowledge of the present, and that is surely wrong.

In the model we draw pictures of things as they might be. An example situation supporting $E \rightarrow F$ does no more than confirm that for some choices of E and some choices of F the claim holds — i.e. it isn't self-contradictory. But you knew that already: there are some Es and Fs for which $E \rightarrow F$ is provable, and in those cases the future must be like an example diagram in the model.

An example situation for $E \rightarrow F$ does no more than ask "what if E and F were chosen so that $E \rightarrow F$ holds, like this?". That's not letting the future influence the present: it's just drawing one possible future. I come back to the notion that a valid claim is a kind of promise: no matter what the future, it will always hold. Proof settles that matter in one direction, disproof in the other. For disproof, you have to display a contrary situation: that means a diagram of a possible future. That's all.

9.3 Notation

9.3.1 Atomic formulae and presence markers. An **atomic formula** is either an identifier, a predicate instance or a relation instance.

An **identifier** is a single formula name — E, F, G or $Goldbach$, for example.

A **predicate instance** is a single predicate name with individual-name arguments — $R(i)$ or $\mathsf{Good}(j)$, for example.

A **relation instance** is a single relation name with individual-name arguments — $T(i, j, k)$ or $\mathsf{Deserves}(i, favour)$, for example.

A **presence marker** is "actual" followed by an individual name — actual i or actual j, for example.

9.3.2 Worlds and situations. A **world** is a blob in a diagram, plus all the atomic formulae and presence markers written next to it.

A **situation** is a world plus any worlds you can reach from it by moving upwards along lines in the diagram, in one or more steps.

The **monotonicity** condition is that when there is a line from world w leading upwards to w', w' must include at least all the atomic formulae and presence markers of w.

9.3.3 Forcing.

$s \Vdash A$

(pronounced "s **forces** A") claims that in situation s you are **forced to accept** formula A. As in Part II, A stands for any formula at all, atomic or composite.

Note that it's $s \Vdash A$, not $w \Vdash A$. Situations force claims, not worlds, even though single worlds with no children are situations, and even though in some cases you can get away with looking at only the root world or a single tip world of s to decide whether or not $s \Vdash A$.

9.3.4 Sub-situations.

$s' < s$

states that s' is a situation you can reach by moving upwards in one or more steps from the root world of s;

$s' \leq s$

states that either s' is s itself, or $s' < s$ (i.e. you get to s' in zero or more steps upwards from the root world of s).

In Fig. 9.2, for example, if s is the entire diagram there are six situations $\leq s$, and five $< s$.

9.3.5 If and only if. The symbol '**iff**' is pronounced '**if and only if**'.

A **iff** B

means that when A holds, so does B, and vice versa. It's a kind of logical equality.

The force of **iff** in a definition A **iff** *def* is that when you have *def* you have A — because of 'if' — and when you don't have *def* you don't have A — because of '**only** if'.

9.3.6 Truth everywhere. $\top$ is forced in every situation, so we don't write it on the diagrams (else we'd have to write it everywhere).

9.3.7 Contradiction nowhere. We aren't allowed to write contradiction on a diagram, so it can't be forced directly, and it's a property of the model that it is then impossible to describe a world at which a contradiction — $E \wedge \neg E$, for example — is forced.

9.3.8 Atomic formulae

Definition 9.1 When A is an atomic formula, $s \Vdash A$ **iff** A is written at every world of s.

Because of the monotonicity condition, if an atomic formula is written at the root world of a situation it must be written at every child world, at every grandchild world, and so on — that is, at every world of the whole situation. So Definition 9.1 is equivalent to

Definition 9.2 When A is an atomic formula, $s \Vdash A$ **iff** A is written at the root world of s.

Definition 9.2 is easier to use: we only have to inspect the root world to check forcing of atomic claims.

None of the situations in Fig. 9.1 force you to accept the atomic claim *Goldbach*, though two of them contain sub-situations that do. In Fig. 8.3(a) on page 125 you are forced to accept the atomic claim F, but not in Fig. 8.3(b). In Fig. 9.2 four sub-situations force E, four force F, one forces G, one forces H, and one (the whole) forces no atomic claims at all.

9.4 Connectives definitions

In formal proofs each step of an argument is written down and justified in full. In Kripke diagrams only the atomic formulae are written down, and not the composite formulae we deduce from them. You have to deduce what the diagram forces by using semantic definitions, and the definitions, like the rules of the logic, cover arbitrarily complicated cases.

You won't be seriously misled if you read the definitions thinking, at first, of atomic formulae — Es and Fs in place of the As and Bs. We'll deal with composite examples later: the mathematics works perfectly for them too.

In Part II the inference rules for $\wedge$ and $\rightarrow$ were relatively straightforward, whereas those for $\vee$ and $\neg$ were more subtle. In semantics $\wedge$ and $\vee$ have straightforward **positive** definitions, while $\neg$ and $\rightarrow$ have trickier **negative** definitions. To use a positive definition you look for supporting evidence; to use a negative definition you look for a lack of opposing evidence.

9.4.1 Conjunction ($\wedge$)

Definition 9.3 $s \Vdash A \wedge B$ **iff** $s \Vdash A$ and also $s \Vdash B$.

$A \wedge B$ means both A and B, right here, right now. To discover if s forces $A \wedge B$, check to see if it forces A and check again to see if it forces B.

9.4.2 Disjunction ($\vee$)

Definition 9.4 $s \Vdash A \vee B$ **iff** $s \Vdash A$ or $s \Vdash B$ or both.

$A \vee B$ means one or the other of A or B, right here, right now. To discover if s forces $A \vee B$, check to see if it forces one or more of A and B.

9.4.3 Negation ($\neg$)

Definition 9.5 $s \Vdash \neg A$ **iff** for every $s' \leq s$, $s' \nVdash A$.

$s \Vdash \neg A$ means that A never happens anywhere in s. To discover if s forces $\neg A$, therefore, you have to look everywhere in s and fail to come across A. Stating the same thing negatively:

Definition 9.6 $s \nVdash \neg A$ **iff** there is an $s' \leq s$ such that $s' \Vdash A$.

Whichever definition we use, negation is a negative connective, one where we search for contrary evidence and are forced to accept only if we can't find any. You have to scan all the sub-situations, including the whole, looking for a rogue which forces A. In the nature of negative search, you can stop as soon as you find a rogue, but otherwise you have to keep looking until you've looked everywhere.

In Fig. 9.2, for example, you have to look everywhere to be sure it forces $\neg$*Goldbach* (it does). In the same figure you can stop partway through if you are looking to see if it forces $\neg F$ (it doesn't: there are four sub-situations which force F, and you can stop as soon as you find one).

Local absence isn't enough $s \nVdash A$ means 'we don't have A in this situation'. That isn't the same as $s \Vdash \neg A$, which means 'we don't have A in this situation or in any sub-situation'. In Fig. 9.1(b), for example, $s \nVdash$ *Goldbach*, because *Goldbach* is an atomic claim which isn't written at the root world. At the same time we have $s \nVdash \neg$*Goldbach*, because there is opposing evidence, a sub-situation which forces *Goldbach*. That situation, therefore, is a counter-example to the classical claim *Goldbach* $\vee \neg$ *Goldbach*: neither of the disjuncts is forced.

9.4.4 Implication ($\rightarrow$). In Chapter 3 I described the oddities of proof with the $\rightarrow$ connective: irrelevant price-of-tomatoes implications (B, therefore $A \rightarrow B$) and useless cunning-uncle implications ($\neg A$, therefore $A \rightarrow B$). To preserve soundness, the model has to be consistent with all the oddities of formal proof, and it is.

Definition 9.7 $s \Vdash A \rightarrow B$ **iff** for every $s' \leq s$, if $s' \Vdash A$, then $s' \Vdash B$.

$A \rightarrow B$ claims that everywhere A is forced, B is forced too. To discover if s forces $A \rightarrow B$, you have to look everywhere in s and fail to find A without B. Stating that negatively:

Definition 9.8 $s \nVdash A \rightarrow B$ **iff** there is an $s' \leq s$ such that $s' \Vdash A$ and $s' \nVdash B$.

What if you can't find an s' that forces A? This is cunning-uncle territory: then you won't find an s' that forces A but doesn't force B. There's an absence of opposing evidence, and $A \rightarrow B$ is forced.

What if B is forced everywhere? This is price-of-tomatoes territory: you certainly won't be able to find a sub-situation in which A is forced and B isn't. There's an absence of opposing evidence, and $A \rightarrow B$ is forced.

9.5 Positively upwards, negatively downwards

The diagrams we draw are monotonic in atomic formulae: if a world includes E then so must its children, and their children, and so on. They're also monotonic in composite formulae: if s forces a composite formula A, then so do its children, and their children, and so on, up to the tips of the diagram. That simplifies checking: if you've found something forced in a situation, it's forced in every sub-situation, i.e. at every point inside that situation. Positive evidence spreads upwards.

Negative evidence also spreads, but it spreads downwards. If s doesn't force A then, just because of monotonicity, neither can any parent situation, or grand-parent, or great-grandparent, and so on, down to the root.

When you are checking a positive connective like $s \Vdash A \wedge B$ you have to check two pieces of evidence: $s \Vdash A$ and $s \Vdash B$. When you are checking $s \Vdash A \rightarrow B$, by contrast, you have at least to check $s' \Vdash A$ at every $s' \leq s$; every time you find $s' \Vdash A$ you have then to check $s' \nVdash B$. Unless there is only one world in s that makes much more work than the positive case.

Because negative evidence propagates downwards, it's often best to work that way. If you can find one sub-situation $s' \leq s$ which doesn't force $A \rightarrow B$, then you know that $s \nVdash A \rightarrow B$. That sub-situation can be anywhere within s, and finding it stops the search. So, for negative connectives, you can start at the smallest and richest sub-situations — the tips of the diagram — and work downwards. If you don't find any negative evidence you haven't lost anything — you had to check everywhere anyway — but if you do, you know that the same negativity holds in all the ancestry of that sub-situation, and that's enough to stop the search.

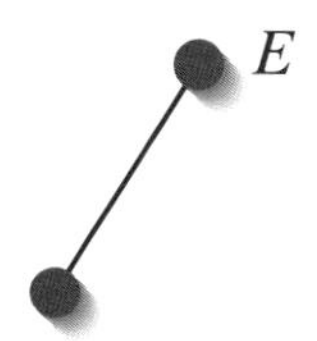

Fig. 9.3 An empty world is not an empty situation

9.6 Empty worlds are tricky

Figures 8.2(c), 8.2(d), 8.3(a) and 8.3(b) on page 125 provide counter-examples to $E \to F \models E$ and therefore, by soundness, to $E \to F \vdash E$, the claim necessary to complete the stuck proof of Fig. 8.1. In each of the counter-examples the root world doesn't mention the conclusion E, because absence is the only way to avoid forcing an atomic claim. In each case the situation forces $E \to F$ because of a lack of negative evidence: in 8.2(c) and 8.2(d) E is never mentioned at all; in 8.3(a) and 8.3(b) every place which includes E also includes F.

The case of Fig. 8.2(d) shows that the single isolated empty world situation — a world with no formulae and no children — is a tricky customer. It forces $E \to F$ because (cunning uncle) it doesn't force E, and that's the whole of the story because there are no other sub-situations. Indeed it forces $A \to B$ for any atomic A and any B at all, so for example it forces $E \to F \to G$. But it doesn't force **every** implication: it doesn't force $(E \to F) \to G$, for example, just because it **does** force $E \to F$ but not G.

The single isolated empty world forces $\neg A$ for any atomic A, for obvious reasons (and then, for the same reasons, it doesn't force $\neg\neg A$).

The single isolated empty world doesn't only force negative formulae: it forces the disjunction $(E \to F) \wedge \neg G \vee H$, for example. It's obvious, though, that it can't force any formula which doesn't contain a negative connective (or quantifier) somewhere.

But — beware! — empty worlds which have children aren't isolated. The isolated empty world forces $E \to F$ and doesn't force $(E \to F) \to G$, for example. Add a world which forces only E (Fig. 9.3) and the picture turns around. Now $E \to F$ is denied everywhere, because we can always reach contrary evidence, and therefore $(E \to F) \to G$ is forced everywhere, even though the situation as a whole still has an empty root world.

9.7 Checking and disproving

There are three distinct activities which are part of disproof:

1. **Checking** a situation to see if a **formula** is **forced** or not.

2. **Checking** a situation to see if a **sequent** is **exemplified** (premises and conclusion forced), **countered** (premises forced, conclusion not forced), or **neither**.
3. **Disproving** a sequent by **inventing** a situation in which its premises are forced and its conclusion isn't.

Chapter 11 deals with the disproof task; I'll concentrate here on checking.

9.7.1 Training wheels for checking. To make the mathematics work, we must only write down atomic claims at a world. That's all fine and minimal, but if we don't write anything else and try to do the calculation in our heads, we can get muddled up. On the other hand, if we do write anything else on the diagram, we have to rub it all out and re-calculate if we change the diagram a bit. Jape is good at that kind of careful book-keeping; humans aren't.

When simply checking a fixed formula against a fixed situation, it's safe to make notes on the diagram. In the examples in this chapter I circle and shade sub-situations to illustrate forcing and to record calculations. You may find it convenient to do the same.

9.7.2 Jape's training wheels. Once you've got a sequent into Jape's disproof pane, you can build situations and Jape will evaluate the claim. If you select a formula in the sequent or text-select a sub-formula, Jape will show you more: it will colour worlds in the diagram violet if your selection is forced there, black if it isn't. You will be able to see for yourself that violet worlds always have violet children, and black worlds always have black parents.

9.8 Checking a propositional formula

Which of the situations in Fig. 9.4 force E? Only 9.4(b) and 9.4(d). Other situations have E, but in the wrong place: an atomic formula is forced only if it is written at the root world of the situation (definition 9.2).

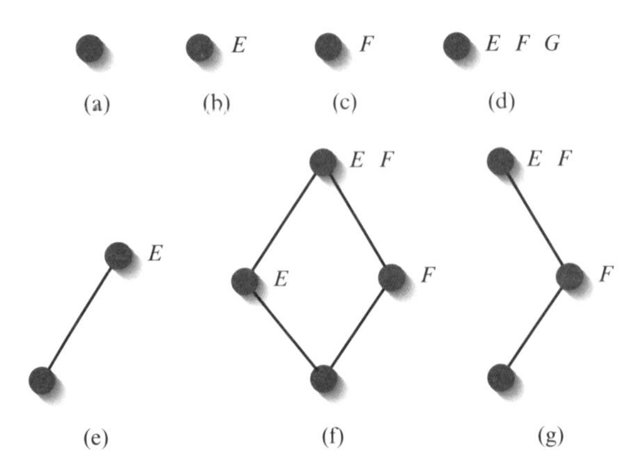

Fig. 9.4 various example situations

Which of the situations in Fig. 9.4 force $E \wedge F$ (Definition 9.3)? Only 9.4(d). Which force $E \vee F$ (Definition 9.4)? 9.4(b), 9.4(c) and 9.4(d).

Positive-only examples are too easy. Which situations force $E \to F$? (Definition 9.8 tells us to look for worlds which force E without F and rule out all enclosing situations.) All of them, except for 9.4(b), 9.4(e) and 9.4(f).

Which of them force $E \to (F \to G)$? More difficult, this: look for worlds which force E without $F \to G$, and rule out all enclosing situations. That means worlds which include E but which can't reach a world that forces F without G. 9.4(f) fails the test at its left-hand and top worlds; 9.4(g) fails it at the top world; all the others pass the test.

Which of them force $(E \to F) \to G$? More difficult still: look for worlds which force $E \to F$ without G, and rule out all enclosing situations. That rules out 9.4(a) and 9.4(c) straight away. 9.4(f), though it doesn't force $E \to F$ overall, has two worlds that do (right-hand and top), and neither mentions G, so it's ruled out. 9.4(g) is ruled out too, because it has three worlds that fail the test.

Which of them force $(E \to F) \to F \to G$? Look for worlds that force $E \to F$ without $F \to G$, and rule out all enclosing situations. 9.4(c) fails; so do 9.4(f) (right, top) and 9.4(g) (everywhere).

9.8.1 Tip: work inside-out in formulae. Complicated formulae make complicated calculations. Semantic definitions, just like proof rules, are expressed outside-in: start with the principal connective and consider its subformulae; look for their principal connective in turn and their subformulae; and so on until eventually you have broken the whole formula down to its atomic parts.

Checking is more efficient if you go the other way round. Information about atomic formulae is written on the diagram. You can use that information to calculate forcing for simple composites, then information about those to make more complicated composites, and eventually arrive at the whole formula. When you get more skilled, you can make short-cuts, but to begin it's better to work systematically from the inside of the formula outwards.

9.8.2 Tip: back your way out of negatives. Nowhere is it more important to work inside-out than when dealing with negative connectives. There is a great deal of experimental evidence that people aren't very good at dealing with negative evidence and contrary reasoning. Looking for counter-evidence for a negative formula can be very confusing, unless you tackle it the right way.

Consider, for example, whether the situation in Fig. 9.5 forces $\neg\neg(E \wedge G)$. According to the definition we must check that every sub-situation $s' \leq s$ is such that $s' \nVdash \neg(E \wedge G)$. That condition contains two negations ($\nVdash$ and $\neg$). If we translate it into a search for contrary evidence — check that there is no sub-situation $s' \leq s$ such that $s' \Vdash \neg(E \wedge G)$ — there are still two negations

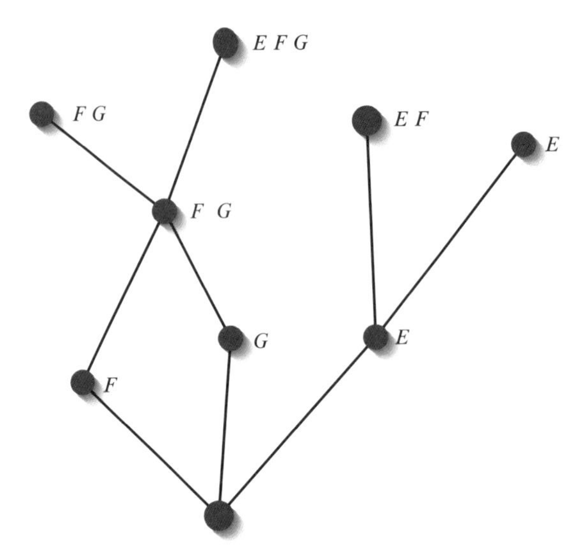

Fig. 9.5 A complicated example situation

('no' and $\neg$). Even to check $s' \Vdash \neg(E \wedge G)$, which contains only one negation, we have to look at every $s'' \leq s$ and make sure that $s'' \nVdash E \wedge G$. But what if we do find a sub-situation where $s'' \Vdash E \wedge G$? Is that evidence or counter-evidence for $s \Vdash \neg\neg(E \wedge G)$? What if we don't find such an s''? Is that evidence, or counter-evidence? I think we are lost in a maze of twisty negations, all alike!

Working inside-out it isn't so hard. Which parts of the diagram force $E \wedge G$? We look everywhere, and find that there's only one place, the tip that forces E, F and G. No situation below that tip can force $\neg(E \wedge G)$; only the leftmost tip and the three-world group top right aren't actually below it, so they must be the only subsituations that force $\neg(E \wedge G)$. That information is summarized in Fig. 9.6. Now we can ask the overall question: does the whole situation force $\neg\neg(E \wedge G)$? Certainly not: there's contrary evidence top left and top right. Job done!

9.8.3 Tip: start negative searches at the tips. The search for evidence against $\neg(E \wedge G)$ above — a search for a place that does force $E \wedge G$ — covered the whole diagram. The contrary evidence was found at a tip. Tips are indeed the most likely place to find contrary evidence, because they have more atomic formulae, but there's an even better reason to start negative searches at the tips: negative evidence expands downwards. If a tip doesn't force A, then neither can its parents nor their parents nor any node below it, all the way to the root of the situation.

Does the situation in Fig. 9.5, for example, force $(G \rightarrow H) \rightarrow E$? Working inside-outwards, we look at $G \rightarrow H$, and the definition tells us to look for places

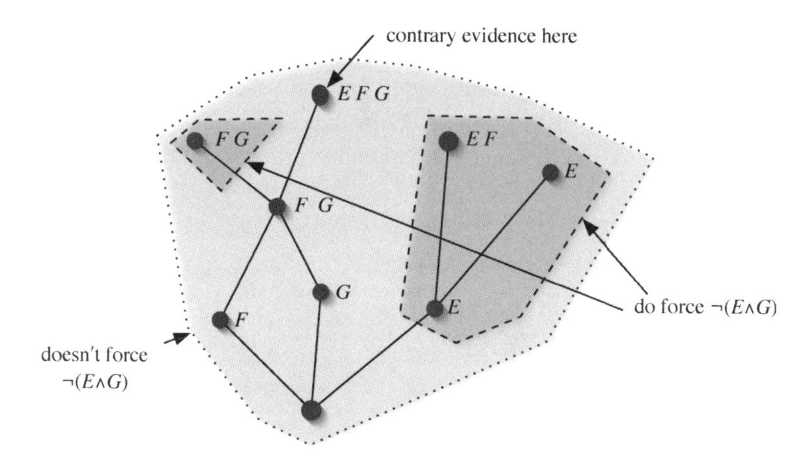

Fig. 9.6 Expanding evidence contrary to $\neg(E \wedge G)$

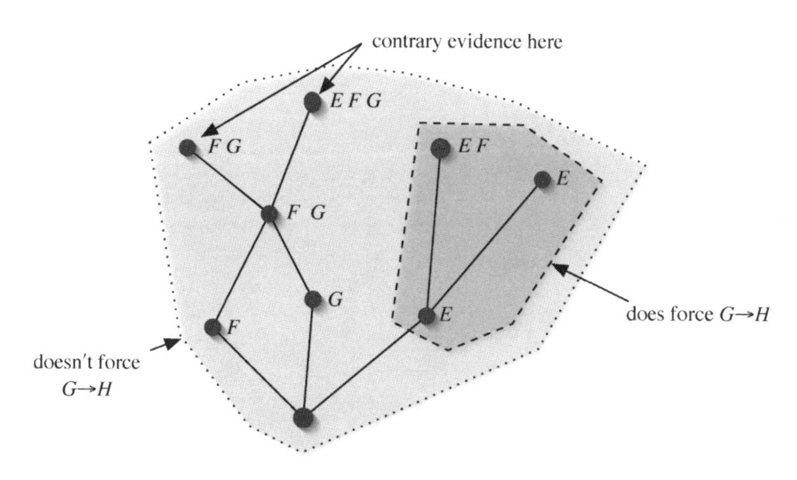

Fig. 9.7 Expanding evidence contrary to $G \to H$

which force G but not H. The two top-left tips (see Fig. 9.7) don't force $G \to H$ (they force G but not H), and that means all the sub-situations on the left side of the diagram, plus the whole situation itself, can't force $G \to H$ either (there is more contrary evidence on that side of the diagram, but we don't need it: the tips are enough).There are no Gs on the right of the diagram, so no possibility of evidence against $G \to H$, so it's forced there, as shown in Fig. 9.7. Now the overall question: does the whole situation force $(G \to H) \to E$? The only possible contrary evidence would be in places that do force $G \to H$ — the triple of worlds top right — and there E is forced too. There's no evidence against $(G \to H) \to E$, so it's forced. Job done!

9.9 Connectives exercises — propositional formulae

Exercise 9.1 Check each of the following formulae against each of the following situations (12 formulae, 9 situations, 108 problems). In each case say whether the formula is forced ($\Vdash$) or not ($\nVdash$) at the root world of the situation:

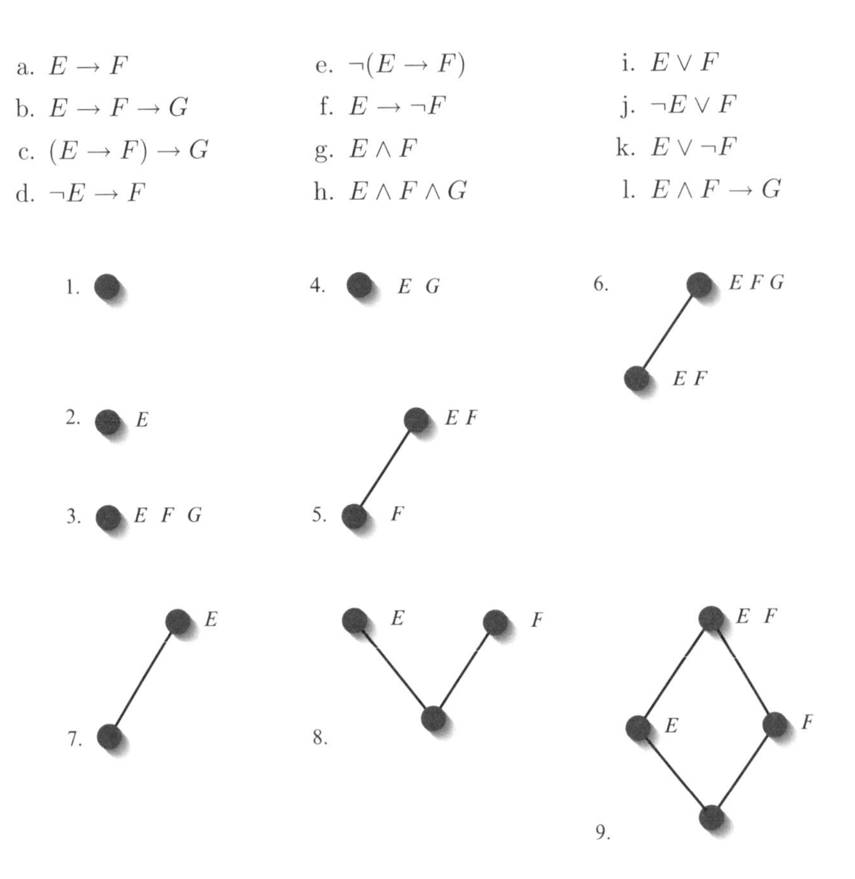

a. $E \to F$
b. $E \to F \to G$
c. $(E \to F) \to G$
d. $\neg E \to F$
e. $\neg(E \to F)$
f. $E \to \neg F$
g. $E \wedge F$
h. $E \wedge F \wedge G$
i. $E \vee F$
j. $\neg E \vee F$
k. $E \vee \neg F$
l. $E \wedge F \to G$

Exercise 9.2 Check each of the following formulae against each of the situations from exercise 9.1 (12 formulae, 9 situations, 108 problems):

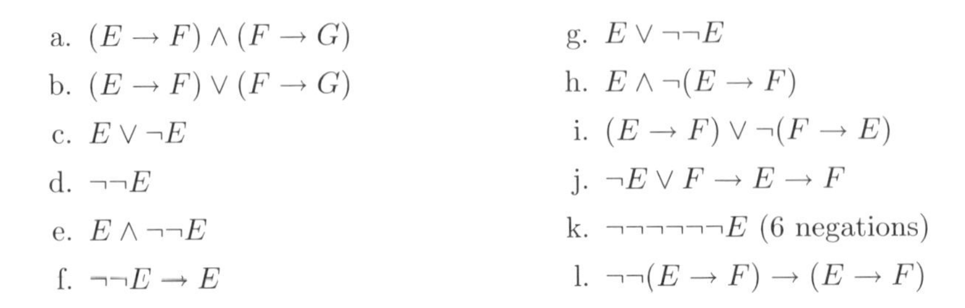

a. $(E \to F) \wedge (F \to G)$
b. $(E \to F) \vee (F \to G)$
c. $E \vee \neg E$
d. $\neg\neg E$
e. $E \wedge \neg\neg E$
f. $\neg\neg E \to E$
g. $E \vee \neg\neg E$
h. $E \wedge \neg(E \to F)$
i. $(E \to F) \vee \neg(F \to E)$
j. $\neg E \vee F \to E \to F$
k. $\neg\neg\neg\neg\neg\neg E$ (6 negations)
l. $\neg\neg(E \to F) \to (E \to F)$

9.10 Checking a sequent

It's pretty easy to check a sequent, once you can check formulae.

- Are **all** the premises forced? (If there are no premises then — guess what? — they're all forced!)
- Is the conclusion forced?

If the answers are "yes, yes", then the situation is an example; if they're "yes, no", then the situation is a counter-example; "no" to the first question means neither example nor counter-example.

9.10.1 Tip: check the easy bits of a sequent first. Suppose you have a lengthy sequent; suppose that some of the premises are pretty complicated, and others look pretty easy to check. Check the easy ones first: if even one of them isn't forced, then there's no need to go further.

Suppose you have a lengthy sequent, a large collection of alternative situations, and you are asked to check which of the situations are counter-examples. Suppose the conclusion looks easier to check than the premises: well then, check it first in each new situation. If it's forced, you don't have to check the premises at all. (The same sort of thing if you are looking for examples, but this time an unforced conclusion resolves the question.)

Exercise 9.3 Check each of the following sequents against each of the situations from Exercise 9.1, saying which (if any) of those situations are counter-examples to the sequent's claim.

a. $\neg E \to F \models \neg F \to E$
b. $\neg E \to \neg F \models F \to E$
c. $E \to F \models \neg E \vee F$
d. $E \to \neg F \models \neg E \vee \neg F$
e. $\neg E \to F \models E \vee F$
f. $\neg E \to \neg F \models E \vee \neg F$
g. $\neg(E \wedge \neg F) \models E \to F$
h. $\neg(\neg E \wedge \neg F) \models \neg E \to F$
i. $\neg(E \to F) \models E \wedge \neg F$
j. $\neg(E \to \neg F) \models E \wedge F$
k. $\neg(\neg E \to \neg F) \models \neg E \wedge F$
l. $\neg(E \vee \neg F) \models \neg E \wedge F$
m. $\neg(\neg E \vee F) \models E \wedge \neg F$
n. $\neg(\neg E \vee \neg F) \models E \wedge F$
o. $\neg(E \wedge F) \models \neg E \vee \neg F$
p. $\neg(E \wedge \neg F) \models \neg E \vee F$
q. $\neg(\neg E \wedge F) \models E \vee \neg F$
r. $\neg(\neg E \wedge \neg F) \models E \vee F$
s. $\neg\neg(E \vee F) \models \neg\neg E \vee \neg\neg F$
t. $\models E \vee \neg E$
u. $\models \neg\neg E \to E$
v. $\models ((E \to F) \to E) \to E$
w. $\models (E \to F) \vee (F \to E)$
x. $\models E \to (F \wedge G) \models (E \to F) \vee (E \to G)$
y. $\neg\neg E \to E \models E \vee \neg E$
z. $\models \neg E \vee \neg\neg E$

Exercise 9.4 Repeat Exercise 9.3, this time saying which situations are examples.

9.11 Quantifier definitions

To define the meaning of quantified formulae in the model, we have to consider population, the universe of quantification, named individuals. That's why the individual-presence markers like 'actual i' can be included in diagrams.

In formal proof the $\forall$ quantifier has simpler inference rules. In the semantics $\exists$ is easier, because it has a positive definition.

You won't go very far wrong if you think at first of atomic predicates whenever you see P in these definitions. Later you can deal with the fact that P stands for any predicate — any formula whatsoever, which may have one or more name-shaped gaps in it.

9.11.1 Individual presence

Definition 9.9 $i@s$ **iff** actual i is written at the root world of s.

(Pronounce $i@s$ as "i at s" or "i is at s".) Because of the monotonicity condition, just as with atomic formulae, if actual i is written at the root world of a situation, it is written at all its worlds.

9.11.2 The meaning of $\exists$

Definition 9.10 $s \Vdash \exists x(P(x))$ **iff** there is an i such that $i@s$ and $s \Vdash P(i)$.

$\exists x(P(x))$ means that you can point to an individual with property P. The constructive definition requires you to point, and can be read as "find an actual i written at the root world of s for which $s \Vdash P(i)$".

To avoid forcing an existential formula — $s \nVdash \exists x(P(x))$ — there must be no individual which has property P. That's often very easy: the isolated empty world has no individuals, so it fails to force $\exists x(P(x))$ for any P at all, atomic or not, negative or positive.

9.11.3 The meaning of $\forall$

Definition 9.11 $s \Vdash \forall x(P(x))$ **iff** for every $s' \leq s$ and for every $i@s'$, $s' \Vdash P(i)$.

Universal quantification is the most semantically negative operator of all. You have to look everywhere, and at every individual you find there, to see if there is opposing evidence; only if you find nothing to the contrary can you conclude $\forall x(P(x))$. It isn't enough even to check the individuals one at a time: you have to check them in every sub-situation in which they occur. Stating that negatively:

Definition 9.12 $s \nVdash \forall x(P(x))$ **iff** there is an $s' \leq s$ and an $i@s'$ such that $s' \nVdash P(i)$.

This means what it says: look at every sub-situation; check every individual at the sub-situation's root to see if it is forced to have property P there; stop if you find an individual that doesn't pass the test, but otherwise keep going.

The isolated empty world is as bizarre as ever: it forces $\forall x(P(x))$ for any P, because when there are no individuals there isn't any opposing evidence. That same minimal situation is a counter-example to $\forall x(R(x)) \models \exists y(R(y))$: the premise is trivially forced, but the conclusion isn't because there is no individual to witness R.

9.12 Checking a quantifier examples

Once you've mastered implication, quantifiers hold no terrors. Existential quantification is as local as disjunction; universal quantification is just like a grand kind of implication. Checking is pretty straightforward.

Exercise 9.5 Check each of the following formulae against each of the following situations (4 formulae, 8 situations, 32 problems). In each case say whether the formula is forced ($\Vdash$) or not ($\nVdash$) at the root world of the situation.

a. $\exists x(R(x) \land S(x))$

b. $\forall x(R(x) \land S(x))$

c. $\forall x(R(x)) \rightarrow \exists y(S(y))$

d. $\forall x(R(x)) \land \forall y(S(y)) \rightarrow \exists z(T(z))$

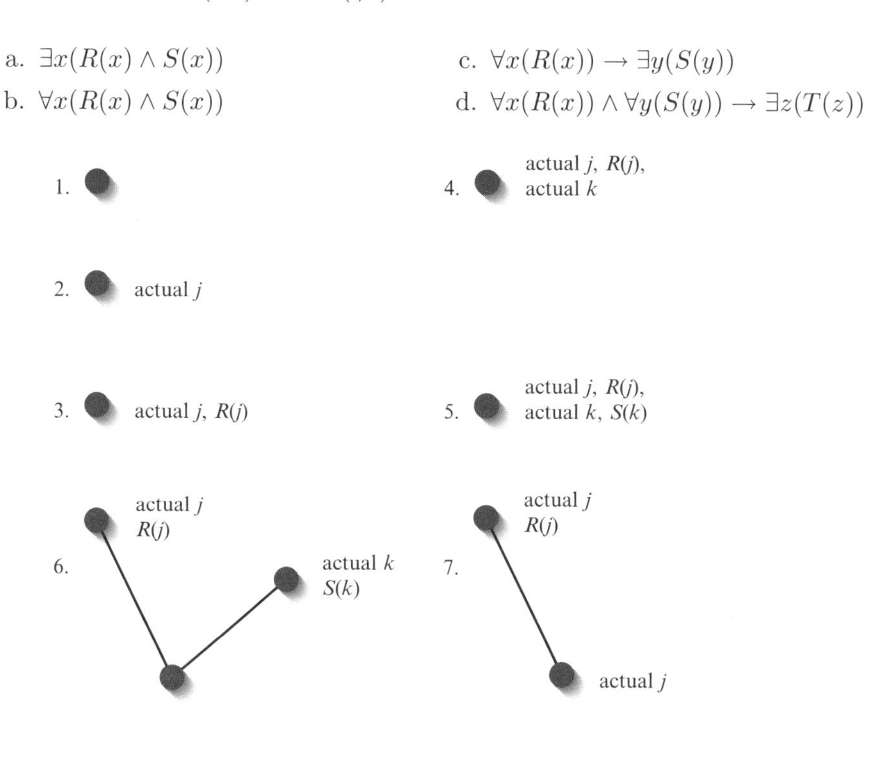

Exercise 9.6 Check each of the following formulae against each of the situations from Exercise 9.5 (32 problems). In each case say whether the formula is forced ($\Vdash$) or not ($\nVdash$) at the root world of the situation.

a. $R(j) \rightarrow \exists x(S(x))$

b. $\exists x(R(x)) \vee \forall y(S(y) \rightarrow T(y))$

c. $\exists x(R(x)) \rightarrow \forall y(S(y) \vee \neg T(y))$

d. $\neg\neg\neg\forall x(R(x))$

9.13 The classical single isolated world

Recall from Chapter 3 that in classical logic a formula makes a claim which is either true or false whether we know it or not. From that point of view it's natural to say that $A \vee \neg A$ is true of any logical formula A. Constructivists, on the other hand, read logical formulae as claims about proof: from their point of view $A \vee \neg A$ is disprovable. Kripke's forcing semantics is designed to explain the constructivist view.

But if we consider only a single world with no descendants, Kripke's mechanism mirrors classical logic. In every situation s either $s \Vdash A$ or $s \nVdash A$, no matter what the form of A; in a single-isolated-world situation, with no sub-situations other than the whole, $s \nVdash A$ is exactly the same as $s \Vdash \neg A$. Excluded middle holds! We have $A \vee \neg A$! The single isolated world describes classical semantics.

If a logical claim has no **classical** proof it will have a classical disproof — i.e. a single-isolated-world counter-example. If it has a classical proof but no **constructive** proof — i.e. if it's a claim in the disputed region of Fig. 3.3 on page 39 — then it can't have a classical disproof, so it will need more than one world to disprove it constructively. Fig. 9.3, for example, is the simplest possible counter-example to the law of excluded middle. The interesting disproofs are the disputed disproofs, all of which need more than one world.

Chapter 11 describes how to find all kinds of disproof, including the interesting ones.

9.14 Contradictory uncle sheep

Contradiction, in logical terms, is a description of confusion. Our formal logic can't live with contradictions: if you could prove A and at the same time $\neg A$ for even one A, you could use $\neg$ intro and either of the contradiction rules to derive any conclusion you like. Since the model and the formal logic ought to precisely correspond, the model must not allow us to describe a situation in which we can be forced to accept A and at the same time forced to accept $\neg A$.

It's clear that the model doesn't support contradiction: if $s \Vdash A$ then $s \nVdash \neg A$, just by the definition of negation. But that means we can't diagram claims that depend on contradiction. It's completely impossible, for example, to construct any situations which either illustrate or disprove $E, \neg E \models F$. The corresponding syntactic sequent is provable, certainly (one step of contradiction, one of $\neg$ elimination), so we should expect to find no disproof. But ... no examples? Can we believe a claim which apparently has no support? Can the model be said to support it?

The cunning uncle of Chapter 3 and the green sheep of Chapter 6 come to our aid. The claim $E, \neg E \models F$ states that when I'm forced to accept E and $\neg E$ at the same time, then I'm forced to accept F: that's a cunning uncle promise if I ever saw one. To prove a claim we must find that every situation which supports the premises supports the conclusion: the green sheep say that since there are no situations which support the premises, all the available situations support the conclusion. (The sheep would be just as happy to work for the prosecution — every available situation is a counter-example, after all — but their accusatory efforts would be in vain, because to make a disproof they'd have to point to a counter-example, and they can't.)

10 Classical semantics

In Part II you saw that constructive proof was easier than classical proof. In Chapter 9 you saw the subtlety of constructive semantics. I dealt with constructive semantics first for two reasons: I wanted to show you that even easy-to-use logic can be **weird**; and I wanted to prepare you to take a sceptical view of the claim that classical semantics is the only model of logic. (I presumed that you would be naturally sceptical of similar claims about constructive semantics: if you aren't, please start doubting now.)

Classical semantics has at least one big advantage over the constructive alternative, and at least two disadvantages. The advantage is that the method of **truth tables** makes calculation with propositional (connectives-only) claims very straightforward, and allows proof **in the model**. The disadvantages are that its explanation of implication is very weak, and, because truth tables only work properly with connectives, many people think that useful and interesting logic stops there. (You already know better.)

You shouldn't be surprised to be told in advance that classical logic is weird too.

10.1 Classical logic and computer science

Computers are machines that deal with very easily decidable questions. A signal, entering or leaving a logic gate, is read as either on or off, 1 or 0, true or false, with no possibility of indecision allowed (it's easier to make hardware if it is only asked to distinguish between two alternatives). Logic gates are then arranged to imitate classical truth-table calculations with 1s and 0s, and combinations of gates imitate the algorithms of al-Khwarīzmī applied to binary numerals.

Hardware memory cells become the variables of our programming languages. So long as we restrict ourselves to remarks about variables and other memory-supported data structures, everything remains decidable. No need to make any proofs to decide if i is or is not 0 — just go and look in the memory cell! — so its quite unexceptionable to insist on excluded middle, and $i = 0 \vee i \neq 0$ really is just **true**.

Not everything is so classically simple. If a proof of a program depends on a proof that a procedure (Java method, C function) does its job, then a proof is what's needed. Excluded middle — "it crashes or it doesn't" — hardly seems relevant.

Still, classical semantics works well for hardware, and for very many questions that we want to ask about our programs. Part IV depends on it. I have to give it a fair crack of the whip.

10.2 Simplicities

In the classical Platonist view, every logical formula is either true or false, independent of our knowledge or our ability to prove or disprove it. The semantics of classical logic, then, has no need of extensions and travel between worlds: everything happens in one place. We can still think of possible worlds — one in which E holds and another in which $\neg E$ holds, say — but we can cover all the possible models of a non-quantified formula with a fixed number of isolated worlds.

If a formula mentions only E there are only two possible worlds: either E is true or it's false.

E
false
true

If it mentions both E and F, there are four:

E	F
false	false
false	true
true	false
true	true

If it mentions E, F and G there are eight:

E	F	G
false	false	false
false	false	true
false	true	false
false	true	true
true	false	false
true	false	true
true	true	false
true	true	true

And so on: in general, for a formula which includes n atomic identifiers and no quantifiers there are 2^n possible worlds. When we've considered them all, we've dealt with all the possible **valuations** of the formula.

Since we only have to consider a fixed number of isolated worlds, we can easily tabulate the logical connectives, treating them as functions on true/false (1/0) values. For example, the classical tabulations ('truth tables') of $\wedge$, $\vee$ and $\neg$:

A	B	$A \wedge B$	$A \vee B$	$\neg A$
false	false	false	false	true
false	true	false	true	true
true	false	false	true	false
true	true	true	true	false

Each row of this table is a A, B situation: in technical language a **valuation**. All possible valuations can be described in just four rows. This makes calculations very simple. In particular, negation, whose semantics is subtle in the rules of Chapter 3 and the definitions of Chapter 9, is just a simple inversion. This is clearly a useful simplification, but it's not a pure advantage: there is a price to pay.

10.3 What's the truth table for implication?

Chapter 3 gave the meaning of $A \rightarrow B$ as "whenever you accept A, you are forced to accept B". In classical terms that means "whenever A is true, so must B be true". It isn't clear how this translates into a truth table, but we can find out by looking at all the possible truth tables for connectives.

In a tabulated A, B situation there are four rows. Each column is an arrangement of four values, each of which is either true or false. Obviously, there are only 16 — 2^4 — possible columns.

Table 10.1 shows every valuation of every possible binary (two-place), unary (one-place) and 0-ary (constant) classical connective, displayed in a sort of numerical order, reading the columns as binary numerals from bottom to top (0=false, 1=true). I've labelled the columns where possible with conjunction, disjunction, negation and equivalence ($\equiv$, same as **iff**) connectives.

The two halves of the table — columns 0–7 above, 8–15 below — are negated mirror-images of each other: 0 is the opposite of 15, 1 of 14, and so on, and vice versa. Most of the columns have obvious meanings, especially when you recognize NOR ($\not\vee$), NAND ($\not\wedge$) and XOR ($\not\equiv$) from the hardware designer's handbook. Only four columns — 2 and 4 in the negative section, 11 and 13 in the positive — are unassigned. The truth table for implication must be one of those four, unless it's to be equivalent to some other connective, which would be very unsatisfactory.

Classical implication requires that "whenever A is true, so must B be true": the bottom two rows of the table are the only ones where A is true, and only in the bottom row is B true as well. That rules out the entire first half of the

Table 10.1 Truth table of all connectives and symbols

A	B	$\bot$	$A \not\vee B$	$?_2$	$\neg A$	$?_4$	$\neg B$	$A \not\equiv B$	$A \not\wedge B$
false	false	false	true	false	true	false	true	false	true
false	true	false	false	true	true	false	false	true	true
true	false	false	false	false	false	true	true	true	true
true	true	false	false	false	false	false	false	false	false

A	B	$A \wedge B$	$A \equiv B$	B	$?_{11}$	A	$?_{13}$	$A \vee B$	$\top$
false	false	false	true	false	true	false	true	false	true
false	true	false	false	true	true	false	false	true	true
true	false	false	false	false	false	true	true	true	true
true	true	true	true	true	true	true	true	true	true

table, every column of which has false in the bottom row; all the columns in the second half have true there, so implication clearly fits in the second half of the table. In the next-to-bottom row A is true and B is false, and that contradicts "whenever A is true, so must B be true": we need false in the next-to-bottom row. Only column 11, of the unassigned columns, passes both the tests.

Now we can fill in all the labels: column 11 is implication ($\rightarrow$); column 13 is reverse implication ($\leftarrow$); column 4 is negated implication ($\not\rightarrow$); and column 2 negated reverse implication ($\not\leftarrow$).

10.4 Is classical implication weird, or what?

Column 11 of Table 10.1 is the only possible truth table for classical implication, and it reads

A	B	$A \rightarrow B$
false	false	true
false	true	true
true	false	false
true	true	true

(10.1)

The true values in the first two rows give novices a lot of trouble. After reading Chapter 9, you should recognize what's going on: this is a **negative definition**! $A \rightarrow B$, in classical as well as constructive semantics, holds unless there is contrary evidence — and the only possible contrary evidence is A true, B false on the third line.

The classical model of implication is just as weird as the constructive version, and in just the same way.

10.5 Checking formulae with truth tables

When you consider formulae which don't use quantifiers, classical truth-table semantics has a marvellous property. You can write down every possible valuation of a formula as a truth table, and you can discover whether it's valid or not without making a formal proof. In constructive logic, by contrast, you can only show disproof in the model: in the other direction it's usually impossible to be sure that you've covered every possibility, so proof has to be in the formal logic.

A **tautology** is a formula which is provable without any assumptions — that is, an A for which $\vdash A$ is provable. Since Natural Deduction is sound and complete with respect to the classical model, a tautology must be true in every possible valuation, an A for which $\models A$ holds. In the case of propositional formulae — formulae which involve only connectives — we can write down every possible valuation, so we can decide whether a propositional formula is a tautology without needing to prove it formally.

$(E \to F) \vee (F \to E)$, for example, is a classical tautology:

E	F	$E \to F$	$F \to E$	$(E \to F) \vee (F \to E)$
false	false	true	true	true
false	true	true	false	true
true	false	false	true	true
true	true	true	true	true

Even though, by soundness, we don't need to make a formal proof, by completeness there must be one. Fig. 10.1 shows a proof made in Jape, and it's the shortest one I know (there's an alternative proof which appeals to a proof of $E \vee \neg E$, but, surprisingly, it's longer). Clearly, proof-by-valuation is sometimes easier than proof-by-the-rules.

10.6 Absurdities

Once you know its truth table, classical implication can be read as a combination of negation and disjunction. The truth table for $\neg A \vee B$ is exactly the same as the one for $A \to B$:

A	B	$\neg A$	$\neg A \vee B$
false	false	true	true
false	true	true	true
true	false	false	false
true	true	false	true

(10.2)

That always seems a shame to me: it is as if implication, which is an attempt to explain an important kind of reasoning, is thrown away.

Looked at another way, it's just the other side of the absurdities that I've been banging on about ever since Chapter 3: the price-of-tomatoes absurdity

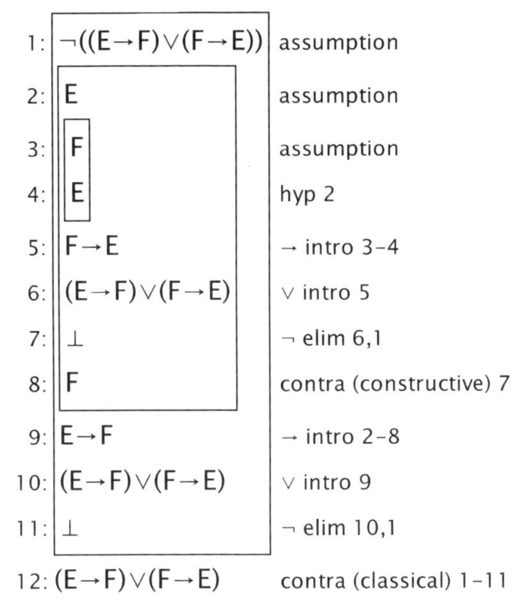

Fig. 10.1 A classical proof of $E \to F \vee F \to E$

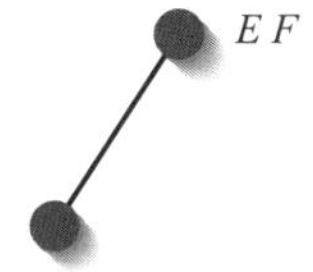

Fig. 10.2 A constructive disproof of $E \to F \models \neg E \vee F$

is $B \vdash A \to B$ and the cunning uncle is $\neg A \vdash A \to B$. Constructivists and classicists agree that $\neg A \vee B \vdash A \to B$. The classical model provides, symmetrically, that $A \to B \vdash \neg A \vee B$ and then, by soundness, it follows that classically $A \to B \vdash \neg A \vee B$. Constructivists don't agree: see the disproof in Fig. 10.2.

Hardware designers don't build logic gates that mimic implication. That's because it's not a good basis for making other truth tables, unlike $\not\vee$ or $\not\wedge$. Most programmers don't even know that implication exists, let alone what its truth table is. That must tell us something about classical implication: simplicity of definition isn't everything.

So classical implication is simple but boring. Constructive implication is a bit more subtle but also a bit tricker. They are both more than a little absurd. All the stuff about promises in Chapter 3 is about the way that the mathematical semantics reflect those subtle absurdities.

You should realize by now that the absurdities of implication aren't a side issue, they are an important part of the semantic definition of Natural Deduction. It isn't easy to unpick the tangle and remove the absurdities or make them less prominent. Logicians have tried to build formal systems in which you can prove $E \to F$ only if the proof somehow calls upon E. They haven't had much mainstream success, though modern developments like linear logic and BI are blazing a trail in computer science circles at least. The phenomenon of the green sheep of Chapter 6 suggest that the cunning uncle will be even harder to banish.

10.6.1 Isn't it better to keep it simple? From the previous discussion you might be tempted to conclude that the semantic differences between constructive and classical logic are a bit overdone. Perhaps it's six of one and half a dozen of the other. Perhaps, in the matter of implication, constructivists swallow an absurd tree-shaped camel and then strain at a silly negation-sized gnat. And surely classical semantics is just **simpler**: no uncles making promises, no possible worlds, just truth tables and useful algebraic equivalences like $\neg\neg E \equiv E$, which constructivists over-fastidiously disdain.

There is a sense in which we can all agree with this argument. In circumstances in which we can reasonably argue that E is either true or false, 1 or 0, on or off, classical logic and its truth tables are just the right thing to do. Hardware logic is a prime example.

There is also a sense in which we ought to disagree. Constructive logic has a famous correspondence with the logic of program-typing, and it gives just the right meaning to calculations in so-called declarative languages, which do without assignment altogether, avoiding just those deductions which, like $A \vee \neg A$, don't appeal to evidence.

Horses for courses, then. In the end the justification for considering constructive logic as well as the classical version is that it gives you a glimpse of a world of mathematics beyond the trivial. It introduces you to the difficulties of precise definition. It doesn't let you suppose that everything is cut and dried; it doesn't leave $\to$ out on a limb; nor does it let you think that $\forall$ and $\exists$ are impossibly difficult. In my own work, in program proving, my colleagues and I invent logics and their models rather often: we have to be prepared to be classical on Monday and constructive on Friday, most weeks.

10.7 Truth tables for quantifiers?

If $(E \to F) \vee (F \to E)$ is a classical tautology, then surely so must $\forall x((R(x) \to S(x)) \vee (S(x) \to R(x)))$ be. The proof is just a matter of introducing an individual j, specializing to $(R(j) \to S(j)) \vee (S(j) \to R(j))$, and then following the steps of Fig. 10.1. That mimics the way the $\forall$ introduction step works.

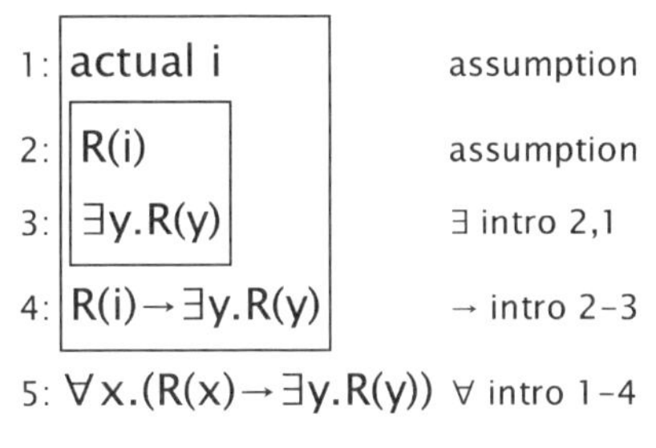

Fig. 10.3 Proof of $\forall x(R(x) \rightarrow \exists y(R(y)))$

Existentials aren't so easy to deal with. $\forall x(R(x) \rightarrow \exists y(R(y)))$, for example, is an obvious tautology: every time you find an individual i with property R, you can be sure that there really is some individual with property R (it's i!). The proof is trivial, shown in Fig. 10.3. $R(i) \rightarrow \exists y(R(y))$ is an obvious tautology, in a universe which includes i, but if we treat the existential like a universal we go wrong: $R(i) \rightarrow R(j)$ is not a tautology at all. The point is that existentials are about choice, and we can't make the choice in advance by inventing an individual: the individuals have to be there first, as the $\exists$ introduction rule makes clear.

The solution to the truth-tables-for-existentials problem is beyond the scope of this book, but if you study logic programming or mechanical theorem proving you will be sure to encounter Skolem functions and discover that the quantifier-free form of $\forall x(R(x) \rightarrow \exists y(R(y)))$ is $R(i) \rightarrow R(f(i))$. Explanation is beyond the scope of this book — but I don't need to explain, because we already have a proof, and in a sound and complete logic that's quite enough.

In classical semantics negated universals are equivalent to existentials, and therefore just as tricky, and it's not easy to deal with universals on the left of implications. A few quantified formulae fit the truth table method, but most don't. I shan't go further into the complexities here. In any case, Chapter 11 shows how to deal with this problem far more directly and convincingly.

10.7.1 Classical semantics of quantification. The classical semantics of universal quantification are just what Chapter 6 say they are: each individual you encounter must have the described property. The method of choosing identifiers and building a truth table imitates the $\forall$ introduction rule: find an individual which is private to the proof of the quantification, and prove the property for that individual; if you can do that then the proof applies to any individual at all.

The classical semantics of existential quantification are also as described in Chapter 6: you must be able to find an individual with the described property. The need to **choose** rather than to invent is what makes the $\exists$ introduction rule difficult to imitate, and what scuppers the truth table 'method' for existentials.

There is a subtlety in the semantics of the existential. Chapter 6, in discussing the universal drunk example, explains an essential difference between the classical and constructive treatments: classically it is enough that there should be an individual with the quantified property; constructively it is essential to point to that individual, to show the witness. This mimics the distinction between the models: classically a formula is true or false whether we know it or not; an existential can be true, even provably true, but we need not say exactly why, nor do we even need to know; constructively, you have to show me.

10.8 Summary

Propositional formulae — formulae made up of identifiers and connectives only — are a pushover for classical semantics. The truth table method eats them up, and can even substitute for formal proof. Add quantification, though, and the picture changes: truth tables stop working and we have to resort to the kind of methods discussed in Chapter 11. The classical treatment of propositional logic is simple and useful, but that doesn't extend to quantifiers.

The cost of the simplicity of classical propositional calculation is great violence done to the meaning of implication, a model so stifling that, in my opinion, it snuffs out the beauty of the corresponding inference rules. The classical existential is simply mysterious, and I don't like that either.

But never mind the aesthetics: classical logic sometimes has its advantages, as the method of truth tables shows in the next chapter, and as Part IV will underline. The sophisticated point of view is that there are many logics, and when we need a logic we pick one to suit our needs. I hope to make you wonder about the use and meaning of logic and perhaps, with Part IV, to tempt some of you into the program-proving woods with me for some midnight logic-bending.

11 Disproof calculation

The proof strategy of Part II is almost a mechanical procedure, in which the discovery of a proof is driven by the shape of the formulae and some simple slogans. Disproof, especially constructive disproof, doesn't seem to be so straightforward. But in fact disproofs can often be calculated.

There are three ways to do it. The first and most mechanical, when it works, is to draw a truth table and read off disproofs by finding rows in which all the premise formulae are forced (are true) and the conclusion formula isn't forced (is false). If there's a classical disproof, and if it's possible to draw a truth table, you'll find it that way.

The second way is to try to make a proof, and fail. A stuck proof has gaps which we don't know how to bridge. If we've used up all the premises and assumptions, worked on each and every connective and quantifier so that they are reduced to atomic formulae, and we've done the same to the conclusions, then we may be able to build a disproof by drawing a diagram which forces the atomic hypotheses we've deduced and doesn't force any atomic conclusions. It isn't quite that simple, and it doesn't always work, but it's very helpful when it does.[1]

After a little practice you will come to the third way: thinking of what you need to force a formula and what you need to deny it. It's tricky, and you at first you may need pencil and lots of paper, or a big blackboard, or a very long reflective ride on the bus, but it's ultimately the most satisfying.

11.1 Simple classical calculations

If a propositional claim (one with no quantifiers) has a classical disproof, then it can be found by truth table, and that's often the easiest way to find it.

11.1.1 $E \vee F \models E \wedge F$. The claim is obvious nonsense — having one or the other of E and F doesn't mean you have both. A two-row truth table is easy to build, as shown in Fig. 11.1. We're looking for a line on which all of the premises are forced (are true) and the conclusion isn't forced (is false). Either line 2 or line 3 will do as a counter-example; line 4 is an example; and line 1 is

[1] This means of searching for disproofs is related to the method of **semantic tableaux**. The presentation in this book doesn't do justice to that elegant method.

E	F	$E \vee F$	$E \wedge F$
false	false	false	false
false	true	true	false
true	false	true	false
true	true	true	true

Fig. 11.1 A truth table which shows that $E \vee F \not\models E \wedge F$

Fig. 11.2 Situations which disprove $E \vee F \models E \wedge F$

irrelevant because the premise isn't forced. Translating lines 2 and 3, I can draw the counter-example diagrams in Fig. 11.2.

We don't really need the tables to point us to the diagrams. We need to build a situation which forces the premise and denies (doesn't force) the conclusion. To deny $E \wedge F$ one or both of E or F must be missing; to support $E \vee F$ one or both of E or F must be present. It's immediately obvious that the worlds of Fig. 11.1 are counter-examples.

It's possible to read the same counter-examples in the stuck proof attempt of Fig. 11.3. There are two gaps in the proof: one asks us to prove from hypotheses $E \vee F$ and E the conclusion F; the other asks us to prove E from $E \vee F$ and F. The worlds of Fig. 11.1 are made by listing the atomic hypotheses at one of the sticking points and not listing the corresponding atomic conclusion — E without F at line 3 and F without E at line 7.

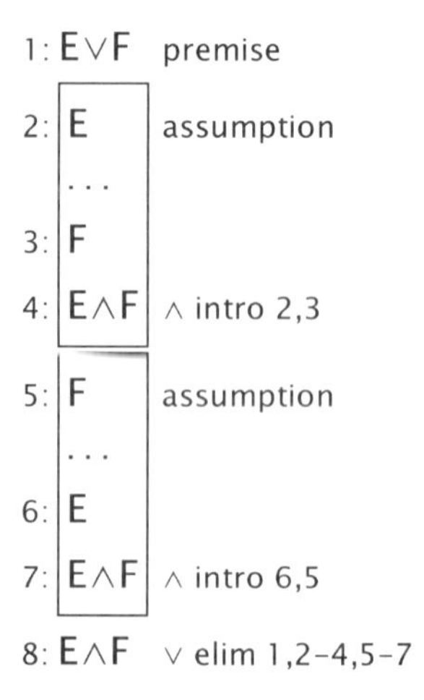

Fig. 11.3 A stuck proof of $E \vee F \vdash E \wedge F$

11.1.2 $E \to F \to G \models (E \to F) \to G$. The claim is not, perhaps, obvious nonsense, especially because in the opposite direction it's easily proved. It isn't valid, though, and the truth table in Fig. 11.4 shows why. In an empty world (line 1) we have $E \to F$ because we don't have E; but we don't have G, so the conclusion is denied; the premise is forced, on the other hand, just because we don't have E. The other counter-example world (line 3) forces just F: we therefore have $E \to F$; we don't have G, so the conclusion is denied; we don't have E, so the premise is forced. (Those are the only counter-examples in the table: on line 7 the conclusion isn't forced, but neither is the premise, so it's neither example nor counter-example.)

The proof attempt of Fig. 11.5(a) hints at the same counter-examples. Don't force G; do force the hypotheses, both of which are trivially forced if we don't force E. But the proof isn't completely stuck yet: it's possible to push it as far as Fig. 11.5(b), which makes it clear that you mustn't force E, so the only choice you have is whether to force F or not.

E	F	G	$E \to F$	$F \to G$	$E \to F \to G$	$(E \to F) \to G$
false	false	false	true	true	true	false
false	false	true	true	true	true	true
false	true	false	true	false	true	false
false	true	true	true	true	true	true
true	false	false	false	true	true	true
true	false	true	false	true	true	true
true	true	false	true	false	false	false
true	true	true	true	true	true	true

Fig. 11.4 A truth table which shows that $E \to F \to G \not\models (E \to F) \to G$

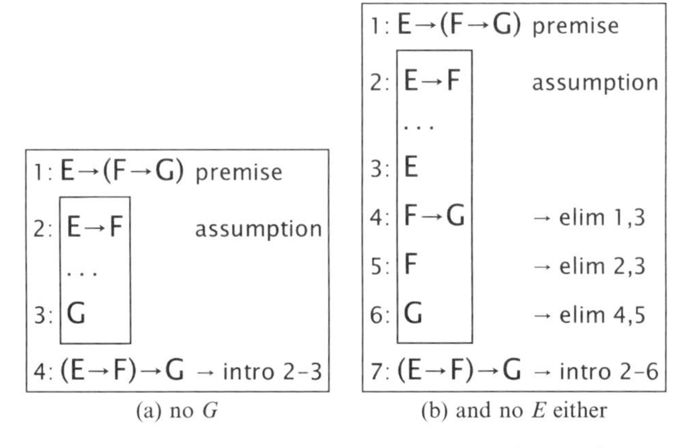

Fig. 11.5 Stuck proofs of $E \to F \to G \vdash (E \to F) \to G$

Without attempting a proof you can reason to the same point. The only way to deny the conclusion is to force $E \rightarrow F$ but not force G. There are several ways of forcing $E \rightarrow F$: not forcing E is one, and that immediately forces the premise, so it's a counter-example. Forcing F is another, but then, since you aren't forcing G, you will fail to force $F \rightarrow G$, so to force the premise $E \rightarrow F \rightarrow G$ you must again fail to force E.

11.1.3 $(E \rightarrow F) \rightarrow G \models E$. This one's obvious: force G, then you have the premise no matter what else; don't force E, and you've avoided the conclusion; F is optional. The truth table is part of Fig. 11.4: lines 2 and 4 are the counter-examples.

In this example the stuck-proof technique doesn't seem so helpful. The statement of the problem tells us immediately not to force E. The obvious first step (Fig. 11.6(a)) suggests that we should force G but not E, but the next step (Fig. 11.6(b)) is pretty hard to read.

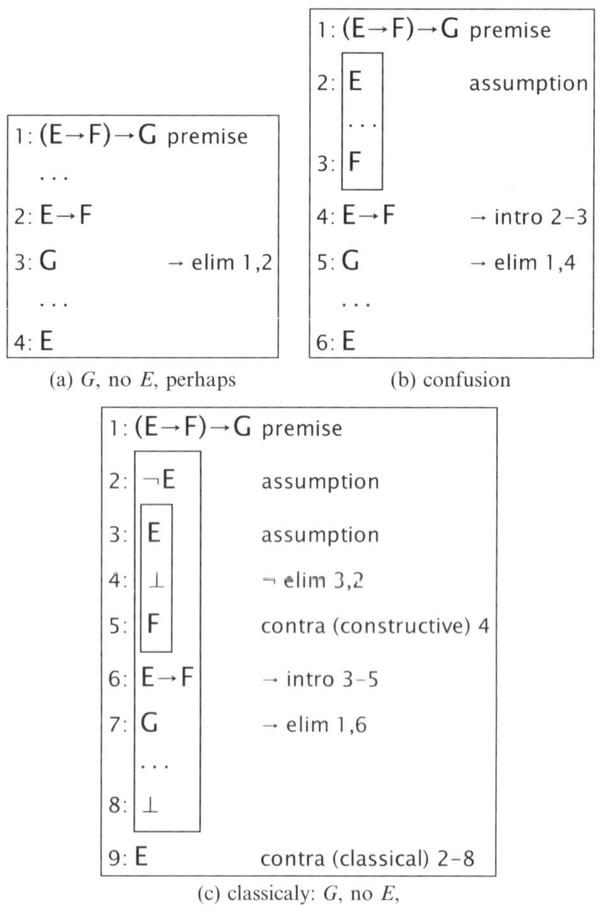

(a) G, no E, perhaps

(b) confusion

(c) classicaly: G, no E,

Fig. 11.6 Stuck proofs of $(E \rightarrow F) \rightarrow G \vdash E$

Classically, though, not forcing E is just forcing $\neg E$, and Fig. 11.6(c) is a totally stuck classical proof with a first step of classical contradiction. It does hint at G without E, but it's a lot of work to find that out given that it was obvious in the first place.

11.2 Classical examples with quantifiers

Quantifiers mess up truth tables, as Chapter 10 explains. Stuck proofs can help, though.

11.2.1 actual $j, \exists x(R(x)) \models R(j)$. This one's nonsense, but it might catch you out. It seems to say that in a universe with only one individual, and a guarantee that some individual has property R, then the named individual must be the one. That particular situation is indeed an example of the claim, but it isn't the claim. The claim is that in **any** universe which includes j and in which somebody has property R, j certainly has it. That's clearly nonsense: all we have to do is to postulate a second individual k who has property R, letting j off the hook as in Fig. 11.7: $j@s$ holds; there's a witness $k@s$ to the existential; j doesn't have property R.

Classically, Fig. 11.7 simply shows $\neg R(j) \wedge R(k)$. You could even draw a two-row truth table, as in Fig. 11.8. But truth tables for quantifiers come after we've found the secret of the disproof. How do we come up with the idea of including $R(k)$?

Stuck proofs are the answer. Fig. 11.9 shows an attempt made in Jape: as soon as we use $\exists$ elim a new individual pops up. Fig. 11.7 uses k rather than i: apart from that, it's a straightforward readout from the presence markers and atomic formulae of lines 1, 2 and 3 of the stuck proof.

11.2.2 $\exists x(R(x)) \models \forall y(R(y))$. Fig. 7.12 on page 113 shows two stuck proofs of this claim. They each tell us the same thing: include two individuals; give one

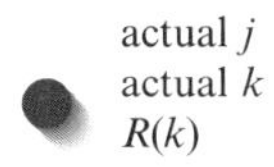

Fig. 11.7 A counter-example to actual $j, \exists x(R(x)) \models R(j)$

$R(k)$	$\exists x(R(x))$	$R(j)$
false	false	false
true	true	false
false	true	true
true	true	true

Fig. 11.8 A truth table for $R(j)$, $R(k)$ and $\exists(R(x))$

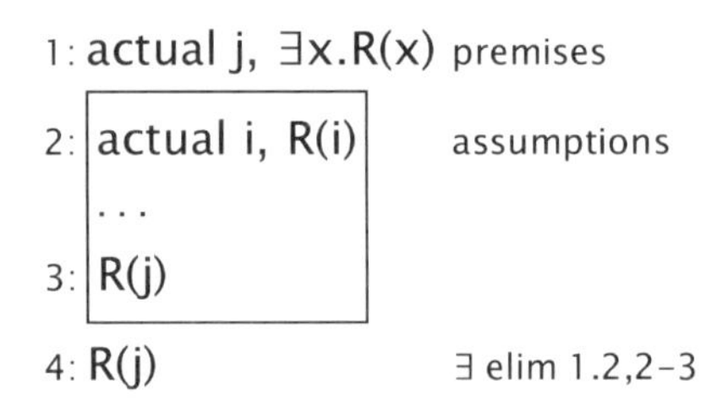

Fig. 11.9 A stuck proof of actual $j, \exists x(R(x)) \vdash R(j)$

property R, but not the other. So Fig. 11.7 is a counter-example. A world with j, k and $R(j)$ but not $R(k)$ would do just as well.

11.2.3 $\forall x(R(x)) \models \exists y(R(y))$**.** A constructive proof of this example can't even begin because there are no individuals to use in $\forall$ elim or $\exists$ intro. You can get started with classical contra, but then you have to show that $\forall x(R(x)), \neg\exists y(R(y)) \vdash \bot$, and you can't.

There are no presence markers and no atomic formulae, so the counter-example is the isolated empty world. Everyone you could possibly meet there has property R; but there is nobody to point to who actually has it.

11.3 Constructive disproof

Classical disproof via truth tables is tedious with large tables, and it's limited as a method. Constructive disproof is more fun, even if you don't want to be a constructivist, because you can examine the edgy claims that classicists accept but constructivists don't.

The only new thing to note is that you have to read stuck proofs differently: boxes can correspond to child worlds. If you always make a child world for each box you may make more than you strictly need, but usually it won't matter very much.

11.3.1 $\vdash ((E \to F) \to E) \to E$**.** This is Peirce's law, beloved of classicists and incomprehensible to constructivists. It's a classical tautology (Fig. 11.10) so it can't have a single-world disproof. All constructive attempts to prove it gets stuck as in Fig. 11.11. The stuck proof at first seems to suggest that a world

E	F	$E \to F$	$(E \to F) \to E$	$((E \to F) \to E) \to E$
false	false	true	false	true
false	true	true	false	true
true	false	false	true	true
true	true	true	true	true

Fig. 11.10 A truth table for Peirce's law, $((E \to F) \to E) \to E$

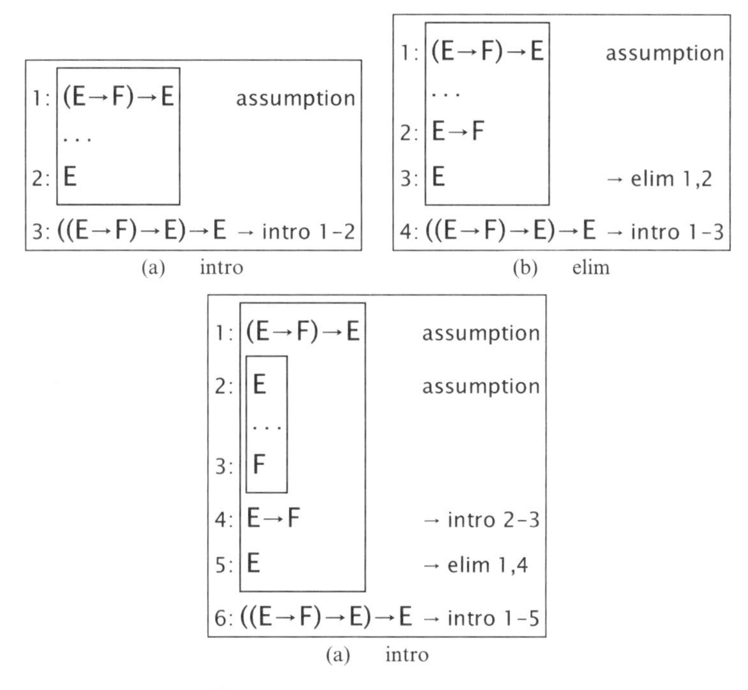

Fig. 11.11 A constructive attempt to prove Peirce's law

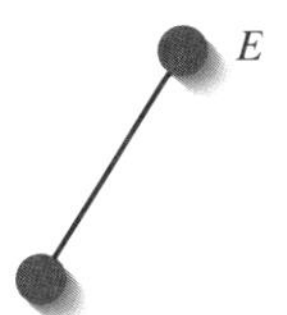

Fig. 11.12 A constructive disproof of Peirce's law and excluded middle

which forces E but not F will be a disproof, but that's not so (Fig. 11.10, line 3). What it actually tells us is to make a child world which forces E and not F, as in Fig. 11.12. $E \rightarrow F$ isn't forced anywhere; so $(E \rightarrow F) \rightarrow E$ is forced everywhere; but we don't have E at the root world, so $((E \rightarrow F) \rightarrow E) \rightarrow E$ isn't forced there. The proof first tells us not to force E; later it says that we must force E, but "inside a box", i.e. in another world. Disproof! Magic!

We don't really need the stuck proof. To deny $((E \rightarrow F) \rightarrow E) \rightarrow E$ we must force $(E \rightarrow F) \rightarrow E$ but not E. Clearly, we must fail to force $E \rightarrow F$, because that's the only way that we can force $(E \rightarrow F) \rightarrow E$ without E. And the only way to fail to force $E \rightarrow F$, as we well know and as the third stage of the proof attempt tells us explicitly, is to have a world that forces E but not

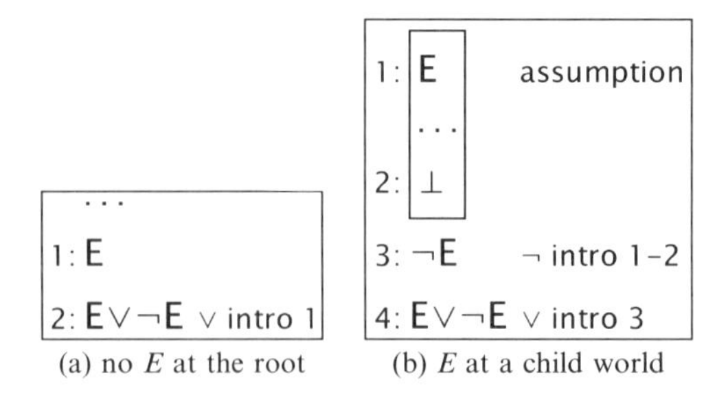

(a) no E at the root (b) E at a child world

Fig. 11.13 Constructivists can't prove excluded middle

F. Since we mustn't force E at the root, we must force it in a child world. Job done; same answer — reflection wins again!

11.3.2 $\models E \vee \neg E$. The law of excluded middle is immediate classically (a two-row truth table — draw it yourself!) but isn't provable constructively. A constructive attempt gets stuck either trying to prove E from no assumptions (Fig. 11.13(a)) or a contradiction from assumption E (Fig. 11.13(b)).

We must show that neither of those proofs is possible: to block a disjunction we have to block **both** sides. The first attempt tells us that we mustn't force E at the root world; the second tells us to have a child world which forces E and make sure we don't get a contradiction (that's easy: we can't make a contradiction in the model anyway). That means that Fig. 11.12 is a constructive counter-example to the law of excluded middle as well as Peirce's law. E isn't forced, and we don't have $\neg E$ either, because we can reach a world which forces E.

11.3.3 $\models (E \rightarrow F) \vee (F \rightarrow E)$. To disprove a disjunction you have to be able to deny each side separately. We have to find a situation which disproves $E \rightarrow F$ and at the same time disproves $F \rightarrow E$. To deny $E \rightarrow F$ we need a world with E but not F; to deny $F \rightarrow E$ we need one with F but not E (we could find that out from stuck proofs, but surely we don't need the crutch). Nothing could be easier: Fig. 11.14 is exactly what's needed. $E \rightarrow F$ isn't forced, because there's a reachable world with E and not F; $F \rightarrow E$ isn't forced either, for similar reasons.

11.3.4 actual j, actual $k \models \exists x(R(x) \rightarrow R(j) \wedge R(k))$. Our old friend the universal drunk has no disjunction, but still has a split disproof. That shouldn't be a surprise: the split is generated by $\exists$ intro, which is after all a generalized form of $\vee$ intro.

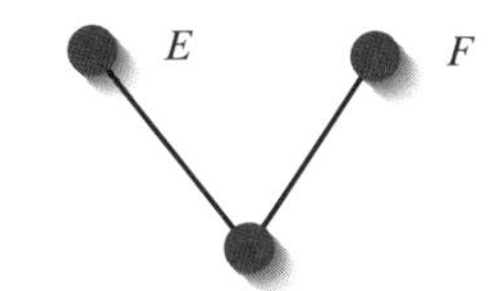

Fig. 11.14 Constructive disproof of $\models (E \rightarrow F) \vee (F \rightarrow E)$

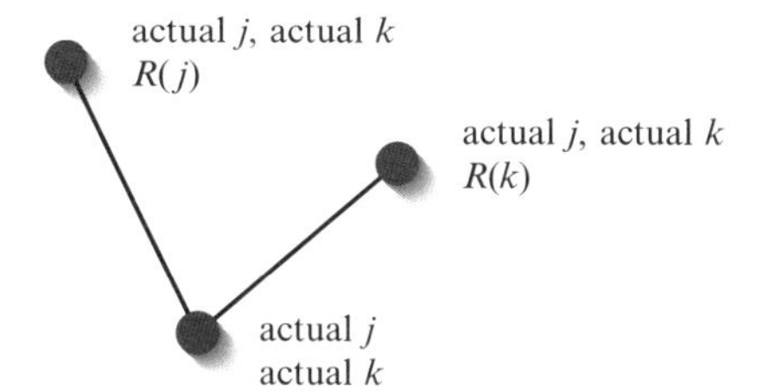

Fig. 11.15 Constructive disproof of the universal drunk

Fig. 7.15 (page 116) shows a stuck proof in which the first step was to specialize the existential using actual j. The proof attempt tells us to make a child world which has $R(j)$ but not $R(k)$. If the first step had used actual k, we'd be told to make a world with $R(k)$ but not $R(j)$. The counter-example, Fig. 11.15, includes both those worlds. Neither j nor k will do as a witness for the conclusion, because each occurs in a sub-situation where $R(j) \wedge R(k)$ isn't forced.

11.4 It isn't always so easy

It might seem that constructive disproof is a breeze, that all you have to do is press the buttons on Jape and read off the answer. Not so!

11.4.1 $\models \neg E \vee \neg\neg E$. In classical logic $\neg\neg E$ is equivalent to E, so classicists read this claim as just another version of excluded middle. Constructively it's quite different, and it doesn't even have the same counter-example as the excluded middle claim.

There's a choice in how to treat the disjunction, and each choice gets stuck, as shown in Fig. 11.16. The left choice tells us to build a world with E in it; the right choice tells us to build a world without E. Fig. 11.17 is the corresponding diagram, and it is the simplest counter-example. $\neg E$ is denied because we can reach a world (top left) that forces E. $\neg\neg E$ is denied because we can reach a

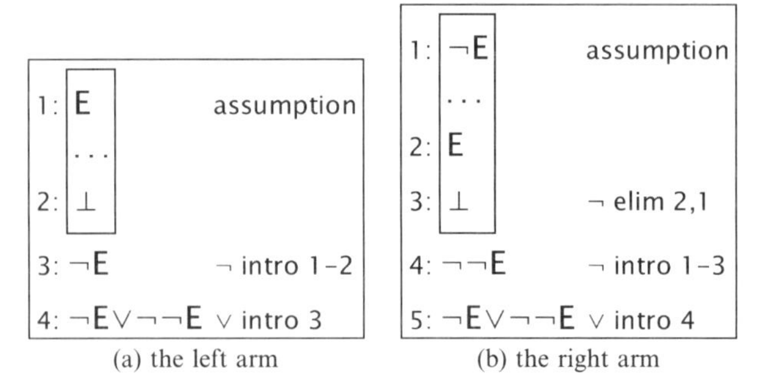

(a) the left arm (b) the right arm

Fig. 11.16 Constructive attempts to prove $\neg E \vee \neg\neg E$

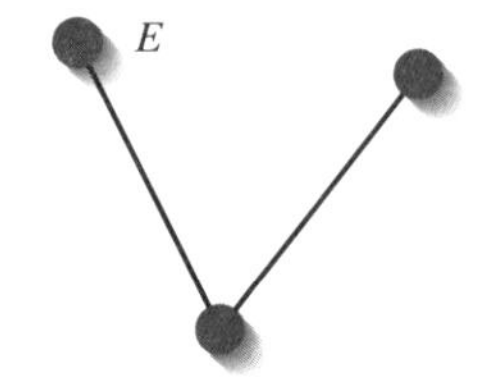

Fig. 11.17 A counter-example to $\models \neg E \vee \neg\neg E$

world (top right) that forces $\neg E$. Fig. 11.12 is enough for excluded middle, but not for this claim.

Adding more negations doesn't make much difference: Fig. 11.17 is a counter-example to $\models \neg\neg E \vee \neg\neg\neg E$. Not all combinations are so complicated: $\models E \vee \neg\neg E$ has a simple classical counter-example, as does $\models \neg E \vee \neg\neg\neg E$.

11.4.2 $\models (\neg\neg E \rightarrow E) \vee \neg E \vee \neg\neg E$**.** Never mind what this monster means: I think it was created especially to bamboozle constructivists. It has a classical proof (try it! build the truth table! only two rows and four columns!) so no single-world disproof. But it has a wonderful constructive disproof.

Constructive proof attempts get stuck in three different ways, as shown in Fig. 11.18. If we read those stuck proofs literally, taking a box as a command to build a child world, and atomic assumptions as the formulae we must force at that world, we'd build a trifurcated diagram like Fig. 11.19(a). That is, indeed, a counter-example. But so is Fig. 11.19(b), made by observing that to deny $\neg E$ all you need is a reachable world which forces E, and there's already one in the left arm, so the middle arm is redundant. Fig. 11.19(b) is simpler. Indeed it is the simplest possible counter-example to the monster claim — which is perplexing because it doesn't **look** simple.

If we are to deny $\neg\neg E$, there must be a reachable world which forces $\neg E$: that's what the empty leaf world is for. If we are to deny $\neg E$, there must be a

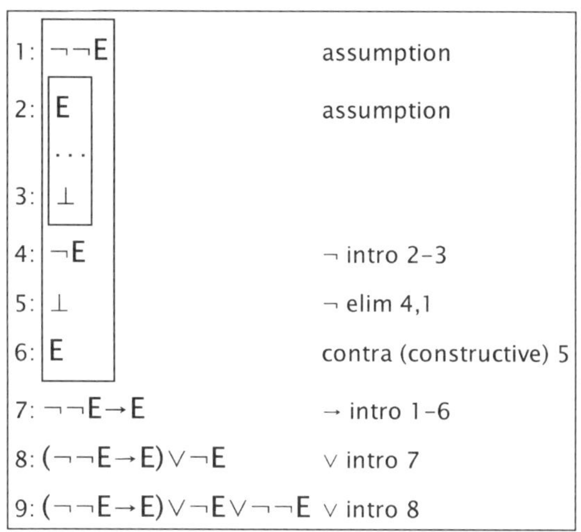

(a) the left arm

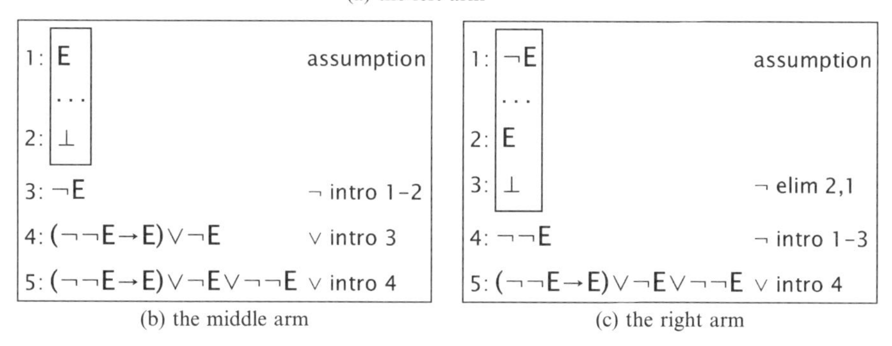

(b) the middle arm

(c) the right arm

Fig. 11.18 Constructive attempts on a monster

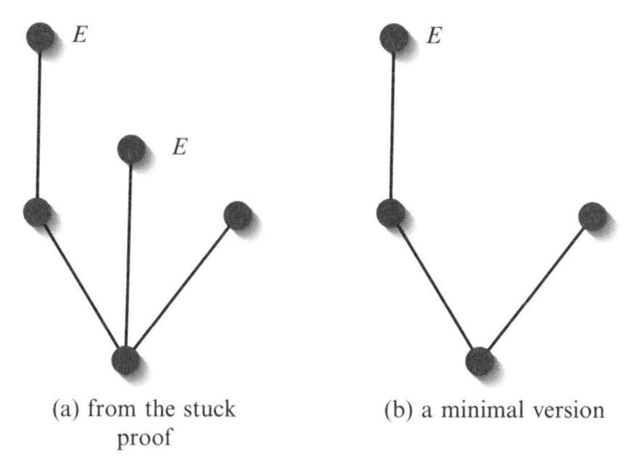

(a) from the stuck proof

(b) a minimal version

Fig. 11.19 Counter-examples to $\models (\neg\neg E \to E) \vee \neg E \vee \neg\neg E$

reachable world which forces E: that's what the other leaf world is for. We still need to deny $\neg\neg E \rightarrow E$: since $\neg\neg E$ is denied at the root and the rightmost tip, and E is forced at the leftmost tip, $\neg\neg E \rightarrow E$ would be trivially forced with just those three worlds. We need a world where $\neg\neg E$ is forced and yet E is not: that's what the intermediate empty world is for.

Part IV

Proof of programs

Logic which merely has a mathematical meaning can seem very dry, almost as pointless as logic without meaning. Logic was invented originally to classify real-world legal arguments, to distinguish good arguments from bad, to provide reasoning with a point. Computer scientists want to be able to write computer programs that work. For us, logic would have a very sharp point indeed if it could help us to program.

This part is about Hoare logic, a treatment of computer programs which wraps classical logic up in yet more logical rules about program statements. The result is remarkable: a logic in which we can claim properties of programs and prove that those properties hold. In particular, we can write programs where the loops do terminate, the array bounds are always respected and the required result is always achieved.

As ever, nothing comes for free. Program proof is tricky and expensive, and most software producers think it's beyond their means. It isn't just a dream, though: it's already one of the ways that the most important safety-critical programs, especially the ones which fly planes and drive trains, are checked before they are used. It remains a dream to extend the range of logic to cover more kinds of program properties, to deal with larger programs, and to make the whole activity easier and more automatic so that everybody can use it. We're working on our dream.

WARNING: the Jape proofs in this part of the book, especially those in Chapters 14 and 15, are large and long. You may be able to comprehend some of them on the page, but you will often find it best to reconstruct them in Jape as you follow the discussion.

12 Specification and verification

The connection between logic and computer science runs through computer programming. Computer programming languages are formal systems — that is, specialized logics. It ought to be possible to exploit that connection and make programs with proven logical properties, programs which don't go wrong or crash as often as programs do today. It's a tempting prospect, and there are at least the following approaches under active development:

- construct a model of the program, and reason within that model (**model checking** and **abstract interpretation**);
- write a specification of the program as a logical claim, and infer the program using logical rules (**refinement**);
- use languages in which program execution is similar to a search for a logical proof (**logic programming**);
- use languages in which program execution is similar to arithmetic calculation, and use equality substitution to reason about program properties (**functional programming**);
- check that an existing program corresponds to a given specification (**verification**).

In this book I discuss only the **verification** of programs written in **imperative** programming languages like Java, C, C++ and the like, because I think it's the most relevant to most of my readers' programming lives.

I would be dishonest if I didn't admit that precise reasoning about programs and their specifications — often called **formal methods** — has been controversial ever since it was first proposed. All sorts of objections are raised against it:

- are programmers clever enough to make proofs?[1]
- what if you get the specification wrong?[2]

[1] This question seems to be posed most often in Anglo-Saxon countries. In England, for example, most schoolchildren learn hardly any mathematics, and it's still acceptable to boast of ignorance of simple algebra. In such an atmosphere, skill with mathematical logic seems almost unnatural.

[2] See Section 14.8. With or without logical intervention, inappropriate specifications are, many researchers believe, the source of a large proportion of bugs in programs. Others think that all specifications eventually become inappropriate, because users' needs change over time.

```
for (c1=buf, c2=s; (*c1++ = *c2++)!=0; );
```

Fig. 12.1 A particularly sharp chisel

- what if the proof machinery is defective?[3]
- what about the programs you can't describe in your logic?[4]
- why is the technique so far used only with safety-critical software (in France, on underground trains and passenger jets)?[5]

This isn't the place to confront those issues. It's enough to say that in a book on logic it's appropriate to look at what logic might have to do with programming.

12.1 A 21st-century embarrassment

Ever since computing became a mass sport, it's been plagued by criminals, vandals and the kind of tricksters who think it's amusing to throw stones at other people's windows. The newspapers call them 'hackers', but since that's a programmers' term of approval, I'd rather call them **crackers**. Crackers find ways round the information protection mechanisms of computers on the internet, mostly for fun but sometimes to vandalize or worse. Even without malicious intent they can do a lot of damage by laying cracked computers open to attack from less casual intruders. What was once just a bit of a nuisance has grown to be a severe pain in the bum.

One of the easiest ways past your computer's defences is via what's called a **buffer overflow**. Fig. 12.1 shows the kind of program which lets in the crackers, written in C with maximum economy and minimum protection to run as fast as possible. What it does is to copy a string `s` into an array `buf`, called a buffer, stopping when it reaches the zero byte which always ends a C string. A similar but less concise way to blow a hole in your own defences is shown in Fig. 12.2, using the notation introduced in Chapters 13, 14 and 15.

[3] This one has an answer. Use a 'proof checker' — a simple program which checks the proofs you prepare. Proof checking is much easier than proof search. Use two checkers, if you don't trust the first one or three, or more.

[4] There are lots of things about mathematics, and particularly (but not especially) the mathematics of computing, that aren't understood yet. If you don't understand something, you can't deal with it logically, or mathematically, or at all. Like everybody else, we find it difficult to deal with floating-point (approximate) arithmetic, with network protocols, and with many other things. We persevere with the things we can do and wait for science to overtake the things we can't.

[5] Well now! See footnote 1.

$i := 0;$
while $s[i] \neq 0$ do $buf[i] := s[i]; i := i + 1$ od;
$buf[i] := 0$

Fig. 12.2 A sharp-enough chisel

The problem — it's the same in either example — is that the string being copied might be too large for the buffer which is supposed to receive it. In those circumstances the C program at least will first fill the buffer and then carry on copying the rest of the string into whatever memory space lies next to that buffer. In carefully researched situations a cracker can arrange the size and contents of the string so that program-control information in memory is overwritten. Choose the right program, the right buffer overflow, the right string, and the cracker is through the defences and in control.

Buffer overflows are only one of our problems. There are lots of other things we need to do to stop crackers, some of them more immediate than the remedies discussed here, but those of us who care about programming have to clean our own house. People are getting in because we are writing programs that don't work as they should.[6] It can't go on.

12.1.1 Should we use blunter tools? C, the language of Fig. 12.1, is dangerous to use because its execution model is very close to that of the underlying hardware. The language level — the logic — describes structures and arrays and strings. In the execution model — the semantics — those linguistic entities correspond to numerical addresses of memory locations, and using address arithmetic (see Chapter 15) a program can do anything it likes with any of those addresses. In practice a C program won't necessarily obey the restrictions which might seem to be explicit in the logic — for example, keeping within the bounds of a buffer array.

Policing execution is one way round the problem. We might require, for example, that the execution checks every indexed access to memory — in Fig. 12.1 `*c1` and `*c2`; in Fig. 12.2 $buf[i]$ and $s[i]$ — to make sure that they stay within the limits of a program-declared array or structure. That's achievable, but it has a cost, because policed programs execute more slowly. Computers nowadays are very fast, though, and almost all of them spend almost all of their time waiting for us to click the mouse or tap the keyboard, so perhaps we can afford the protection.

[6] As I wrote, a buffer overflow exploit was discovered in `rtf2latex2e`, an open-source program that I worked on in order to produce this book. Oh, calamity! (It wasn't in the bit I worked on, though. Phew!)

On the other hand there's a strong sense in which C and other unpoliced programming languages are wrongly accused. First, some programs do have to run as fast as possible: device drivers, network handlers, operating systems in general, programs which draw stuff on the screen, and so on through a very long list of examples. Second, if it is used in the right circumstances — that is, when the buffer actually is large enough — Fig. 12.1 behaves perfectly as well as speedily. The behaviour of the program is described by an implication: **if** the array `buf` is large enough **then** the string `s` will be correctly copied into it. You might wish for more, but a contextual promise is all that any program can live up to.

12.2 Specification as implication

A specification says what an artefact — a building, a washing machine, a program, an insurance company — **ought** to do; the artefact does what it actually **does**. The problem is to bring the two together. Usually, the specification is seen as primary, and the problem is to check that the artefact meets, fulfils, **satisfies** its specification. In the case of a computer program, the problem is to bring an action in programming-language logic in line with a claim in a specification logic.

A specification can be seen as a contract or a promise: if you provide my program with such and such an input, I guarantee that it will produce such and such an output. That's very like a maker's guarantee. When you buy a washing machine, for example, the maker promises that the machine will wash effectively if you

1. plumb it into the water supply and a drain;
2. plug it into an electrical outlet;
3. load not too many clothes;
4. add the right amount of detergent;
5. close the door;
6. press start.

To benefit from the guarantee you have to keep within the bounds of the promise. Miss any steps or make them in the wrong order, and the machine might not start, or it might start but not wash properly.

Often a machine will be capable of more than the manufacturer guarantees. A washing machine will often work quite well, for example, if you use only half the amount of detergent that the maker recommends, and it will wash after a fashion even if you put in none at all. You may find that your cat enjoys sleeping on top of your machine while it's running, or that the leaps and bounds of an overloaded drum crack the floor and reveal an ancient hoard of gold coins.

All of this is very reminiscent of the meaning of implication. Washing machines which work with no detergent replay the price-of-tomatoes argument. Those which are guaranteed to work only if you use one special difficult-to-obtain washing powder run close to the cunning-uncle position.

12.3 Hoare triples

Our program specifications have three parts and are named for C.A.R. Hoare, who invented them. We say

> **precondition:** in what circumstances our program is guaranteed to work;
> **postcondition:** what effect it will have in those circumstances;
> **program:** what program it is that we guarantee.

The Hoare triple

$$\{A\}\, prog\, \{B\}$$

claims "program *prog*, given circumstances satisfying precondition A, is guaranteed to finish in circumstances satisfying postcondition B". Just what a program is, just what it means to satisfy a postcondition and a precondition, we shall see.

It's only a logical guarantee, of course: if the precondition is $\perp$ no circumstances can satisfy it and the specification is cunning-uncle useless. Most of the time, though, a Hoare-triple's promise is worth having: if *prog* is guaranteed to finish, it's guaranteed not to crash and not to loop for ever.

12.3.1 More backward reasoning. Proofs in Natural Deduction are often discovered backwards but afterwards read forwards, and it's the same with programs and Hoare triples. We have a program, we know the effect that we want it to have — i.e. its postcondition — and from that we can often calculate a precondition. If the implication you construct — given this program and that precondition, these are the effects you can rely on — isn't what you want then you have to rebuild the program and/or the postcondition and try again. That's programming — but in this book I'm only concerned with calculation and deduction.

12.3.2 Memory states. Because we are dealing with imperative programs, which work by making changes to data held in a computer's memory, it's natural that our specifications will describe changes in memory content. Given a suitable configuration of the memory to start with — for example

> n is greater than 0, the array a contains n integers already sorted into order, and j contains an integer that you choose

— you might promise that your program will produce the configuration

> n, a and j will be unchanged, and i will point to the item in a whose value is closest to j

That is, you're selling a search program.
Or you might promise that given a memory configuration

> n is at least 0 and array a contains an n-integer sequence Ka that you choose

your program will produce the configuration

> n will be unchanged and a will contain a rearrangement of Ka in non-descending order.

That is, you've written a sorting program.

Computer memories are huge things. The humblest laptop's memory nowadays contains hundreds of millions of memory components. To have to specify the state of the whole memory to describe the smallest program would be absurd. In practice, our programs each work on a small section of memory: we'll specify what should be in that section when they start, and what will be in the same section when they've finished. We shall do so using formulae like $i = 3$, $j < 7$, $u \times v \geq 2^{n-1}$, using program variables and conventional number-algebraic connectives.[7] We'll combine those number-algebraic components into larger claims using the logic-algebra connectives and quantifiers of Natural Deduction: $\wedge$, $\vee$, $\neg$, $\rightarrow$, $\forall$ and $\exists$.

12.3.3 We use classical logic. Program variables are very simple containers of value. To check that $i = 3$ you look into the variable i and see what it contains. It ought to be the binary bitstring 0...011 (all zeros except for two final ones). If that is what you see, then $i = 3$. If not, not. It always is or is not the case that $i = 3$: we can never be in doubt because of lack of proof or disproof. We know, furthermore, a version of excluded middle: $i = 3$ is exactly the negation of $i \neq 3$, and one of the two will always hold. Similarly, $i < 3 \vee i = 3 \vee i > 3$: in every situation exactly one of those tests will succeed and the others will fail.

Don't suppose that these classical properties are obvious and beyond debate. Given arbitrary formulae A and B describing arbitrary calculations it can be impossible to prove $A = B$ or $A \neq B$, and the same objection to excluded middle that was raised in Chapter 3 rears its head. **But**, if we restrict ourselves

[7] If you're used to programming in Java, or C, or Fortran, you will need to be reminded that the symbol '=', pronounced 'equals', is the number-algebra sign for equality. See Chapter 13 for more about the oddity of languages which have it otherwise.

to questions about the contents of variables that we can look into, then the objection doesn't arise, and we can use classical arithmetic and classical logic.

Most of the time the classical / constructive distinction won't be an issue, but when it's necessary to say that the memory is either like this or it isn't, then we can call on excluded middle without a shadow of doubt. With the exception of some difficulties with finite-width arithmetic (see Section 13.1.6), this makes Hoare-logic reasoning relatively simple (though we won't have much use for truth tables, so not **that** simple).

12.3.4 Vagueness of specification: sets of states. Our specifications wouldn't often be useful if each specification could only describe a single memory state. Useful computer programs, like useful washing machines, work in a variety of circumstances and produce results according the situation. You are already used to some vagueness in logical formulae: $E \vee F$ can be forced in more than one way, as can $E \rightarrow F$, and so on. The number-algebra parts of a specification bring even more vagueness: $j < 7$ can be satisfied in, and thus describes, an infinite number of states, because there are an infinite number of integers less than 7.

This gives a way of reading a formula like $0 \leq i < 17 \wedge a[i] \geq 0$. We can see if a particular computer's state at a particular instant satisfies (forces) this formula by looking inside its memory — but that's only checking an example. We read the formula as a claim that the variable i has one of the values 0, 1, $2, \ldots, 16$ and, whatever one of those values it has, the corresponding element of array a is non-negative. That describes a **set** of states: all those which satisfy the specification, with all the different possible values of i and $a[i]$. The set is larger than you think, because you have to allow for the values of other variables as well. There's an infinite set of states in which $i = 1$ and $a[1] = 30$, for example: one in which $j = 1$, one in which $j = 2$, and so on for ever, never mind the value of $a[i-1]$ or the contents of an entirely separate array b.

A nice consequence of this reading is that implication becomes set inclusion (subset, $\subseteq$), and then $A \rightarrow B$ means that the set of states which satisfy A is included in the set which satisfy B.

There's also a nasty consequence: we can avoid stating all the things that our programs do. Washing-machine manufacturers, to go back to my earlier example, are usually careful to state in the small print but not to emphasize the drawbacks of using their machinery. Washers use electrical energy (costs money), water and soap (ditto), make a noise (degrades the environment) and slowly wear out your clothes. Those costs are really part of the guarantee. Most of us, on balance, prefer to pay the small-print costs because of the benefits of the rest of the deal.

Hoare logic needs its own small print. Unfortunately, a Hoare triple **can't** say "and that's all". You might guarantee, for example, only that your program

Walk along looking at the ground. Each time you see a dollar bill, pick it up. Stop when you have a million of them.

Fig. 12.3 McCarthy's make-a-million program

doesn't change a variable j. A program which increments the value in variable i and does nothing else fits that specification — it doesn't change j, after all! The "i doesn't change" specification logically fits **any** program that doesn't change i, and that's surely misleading. A program that starts a nuclear war would fit it, provided only that it doesn't mess around with j. Weaselly mis-specifications like that can seriously mislead.

We can stay honest if our specifications always mention all the variables a program depends on, and certainly all the ones that it changes. A copper-bottomed solution seems to call for a resource logic, which is my current line of research and far beyond the scope of this book.

12.4 Termination

Because the relationship between specification and program is an implication, there is a particular issue which we have to address.

Computer programs, as any programmer knows, sometimes **crash**, attempting an action which is outside the execution model of the programming language. They sometimes **infinitely loop**, apparently doing some internal calculation but producing no visible effects. Programs which don't crash and which don't infinitely loop must eventually come to an orderly halt delivering a result. They are then said to have **terminated**, as a rail journey terminates when the train comes safely to rest at its destination. Hoare triples, in this book, are about programs which terminate.

Some specifications can be written in terms of what is called **partial correctness**, statements about the effect of a program **if** it ever terminates. For my purposes that's an implication too far. John McCarthy, the great 20th-century computer scientist, invented a joke illustration, a version of which is shown in Fig. 12.3. This program, he pointed out, will make you a millionaire **if** it ever terminates. But even those of us who are running the program (I'm better at finding money in the street than anyone I know!) realize that it's very unlikely ever to make us rich. Partially correct programs can flatter to deceive.

On the other hand, some programs — operating systems, for example — are designed to loop for ever and never terminate. It is possible to treat such programs logically, but I don't address that problem in this book.

12.5 Only an introduction

Computer science has enormous achievements to its credit, but it's still a very young subject. It's still small enough for a novice to be taken very quickly very close to the frontiers of knowledge. The problems of specifying and verifying programs are active areas of research and will be controversial for the foreseeable future. There are all sorts of things that can be done, but there are far more things that can't. I'd be dishonest if I didn't emphasize that point right up front.

I'd feel just as dishonest if I didn't try to convey some of the excitement that I and others feel about the development of this field, and show you some of the ground that you may one day decide to explore. I'll even point out the potholes, so you'll know where you are when you fall into one.

13 A simple programming language

Real-world programming languages are complicated, partly because they are designed to support much more than the basic description of machine activity. Modern languages ease the task of maintaining programs, using libraries, explaining programs to managers, and lots more. They have features — variable typing, for example — which are designed to help you avoid simple programming mistakes.

All of that's jolly fine and really useful, but it complicates reasoning. I simplify my task by concentrating on the core of an imperative programming language, the engine room that does the business. I shall leave out a lot of useful features: in particular, I won't be dealing with declarations or types, because all the variables I need will always be around for me to use, and I'll be working only with integers and Booleans. I'll ignore arrays until Chapter 15, and I won't be dealing with procedures (aka functions, methods) at all.

13.1 Basics

The language I describe is an abstraction from imperative programming languages: those which, like Java, C, C++, Fortran, Ada and so on, work by testing and altering values stored in a computer's memory. It isn't a declarative programming language like Miranda or Haskell; it isn't a logic-programming language like Prolog. Those other kinds of languages have their particular forms of reasoning. I choose to deal with an imperative language with Hoare logic because I think it's closer to most of my readers' programming experience.

The whole language is shown in Table 13.1. Most of it is defined in this chapter but loops (while ...) are defined in Chapter 14 and array element assignment ($a[E]$:=...) in Chapter 15. Apart from some minor notational variations, it is

Table 13.1 A little programming language

Instruction	Effect
Skip	Null action (a kind of zero)
$x := E$	**Assignment** to a **variable**
$a[E] := E'$	**Assignment** to an **array element**
prog1 ; *prog2*	**Sequence** of actions
if E then *prog1* else *prog2* fi	Conditional **choice** of action
while E do *prog* od	Conditional repetition of action (**loop**)

the language to which Hoare logic is normally applied. But it isn't exactly like any language in current use, and that fact demands an explanation.

13.1.1 A squabble about assignment notation. The symbol '=', pronounced 'equals', has a long history in school arithmetic and algebra. It's conventionally used for comparison ($x = 3$: is x equal to 3?), specification ($x = 3$: x is equal to 3), and definition ($f(x, y) = x \times y + 7$: f is the multiply-and-add-seven function).

But then, in the early development of programming languages, the designers of Fortran hijacked the equals sign and used it for assignment. Assignment is the command "store this value there!", a memory-changing command and the reason that imperative languages are called imperative. Because equals had been hijacked by assignment, Fortran programmers had to use '`.EQ.`' for the normal purpose of testing equality. This was hardly satisfactory: Fortran was a **For**mula **tran**slator, and here it was breaking the basic formula rules.

The designers of Algol 60, five or six years later, avoided the problem by using ':=' for the novel operation of assignment and reserving '=' for the traditional notion of equality, calming the situation more than somewhat. Programming life was then fine for about 15 years till the designers of C went back to the Fortran well, choosing '=' for 'becomes' and '==' for 'equals'. This has been a nuisance ever since, especially because, unlike Fortran, C's '=' and '==' are algebraic operators. In C you can write '=' where you mean 'equals', and '==' where you mean 'becomes', the compiler won't complain, and you can muddle yourself up pretty well.[1] It was a very very very deplorable decision for which the designers of C — and those of C++ and Java who followed them — should surely do penance.

I've had to take a position in this dispute, and I've gone for the algebraic choice, using ':=' for 'becomes' and the standard number-algebra sign '=' for 'equals'.

13.1.2 Types. To simplify things, I've ignored variable typing. My programs manipulate integers, and there is no data structuring beyond arrays (which must wait till Chapter 15). But already that makes two different kinds of value which a name might refer to, and there has to be a third, because the choice-formulae in if and while commands must evaluate to $\top$ or $\bot$, what most programming languages call **Boolean** values after George Boole who invented Boolean algebra in the 19th century.

So there really are two basic value types — Booleans and integers — plus arrays which can make structures. But I'll be dealing with very small programs, I'll be using helpful conventions (i, j, k for integer variables and a, b, c for arrays) and it'll be easy to get by without explicit typing.

[1] Don't spot the problem? Try reading that sentence aloud.

13.1.3 Punctuation. Semicolon (;) is a separator in English punctuation. The designers of Algol 60 used it as such. C's designers converted it to a kind of terminator (another penance called for, perhaps?).

Most programming languages have some sort of bracketing: for example, Algol's begin, C and Java's $\{\ldots\}$. Brackets are needed because of language constructs which include opening brackets without a matching closing bracket, most often in loop and choice commands. This notational choice makes programs harder to read.

In my language semicolon is a separator in a sequence of commands; conditionals always have two branches; conditionals and loops have opening and closing brackets.

13.1.4 Value-formula notation. Formulae (E, F) in the language can use conventional number-algebra notation and the $\wedge$, $\vee$ and $\neg$ connectives of Natural Deduction. I use $\top$ and $\bot$ where conventional languages use true and false, for compatibility with the logic of parts II and III.

13.1.5 An action-zero. You've been battered by earlier chapters to accept that the empty universe of quantification is, like arithmetic zero, a worthwhile case to consider. In programming the null action is a kind of zero. Most programming languages, unfortunately, indicate it by an absence (there are two such absences in Fig. 12.1, for example). That's like numerical notation before the Hindus intervened. My language has skip as a mark for null action, the zero of programming.

13.1.6 Finite-width-numeral arithmetic. A well-known way to crash a program is to ask for division by zero: the calculation $j \div 0$, for example, doesn't terminate in most execution systems (division by zero would require an infinite loop, and the machine couldn't represent the answer anyway, so both the calculation and the program that called for it have to be abandoned). Another program-crashing trick is to exceed the range of values that can be held in an integer variable: $i := i + 1$, for example, can't terminate if i already contains $MAXINT$.[2]

In arithmetic reasoning, outside the computer, $j \div 0$ is perfectly well defined — it's infinite, the particular infinity called $\aleph_0$ — and $i + 1$ is always defined no matter how large i might be. On the other hand, allowing infinity into arithmetic makes strange things happen: $i = i + 1 = i - 1$, for example, if i is infinite. We

[2] $MAXINT$ is $2^{63} - 1$ in most imperative programming languages on a present-day desktop PC — a big number but not out of reach. If $i := i + 1$ were to terminate when i already contains $MAXINT$ it couldn't possibly have the right effect. Let's hope it doesn't terminate on your machine: let's hope it crashes instead.

ought perhaps to be a little frightened of infinity. We definitely should be very frightened of trying to formalize arithmetic, because of Gödel's proof that it's impossible to do it properly.

But my concern is to specify and verify what programs actually do, not to capture all of arithmetic. Finite-width-numeral arithmetic, which is what machines do and therefore all that an imperative-language programmer can rely on, is so limited that we needn't be frightened of infinity or of Gödel's tricks. If we take the judgement 'E computes' to mean that E can be calculated in finite-width-numeral arithmetic, it's easy to generate a system of rules such as

$$\frac{\begin{matrix}\vdots & & \vdots & & \vdots \\ A \text{ computes} & & B \text{ computes} & & \mathit{MININT} \leq A + B \leq \mathit{MAXINT}\end{matrix}}{A + B \text{ computes}}$$

and

$$\frac{\begin{matrix}\vdots & & \vdots & & \vdots \\ A \text{ computes} & & B \text{ computes} & & B \neq 0\end{matrix}}{A \div B \text{ computes}}$$

In an industrial-strength theorem-prover that's more or less how it's done. In a book like this one, designed to help you learn how logics work and to show you the intellectual difficulties of proof, it would be tedious and distracting to have to demonstrate finite-width-numeral computability at every step of every proof.

Finite-width arithmetic problems are an example of the problem of **definedness**. The other example which I highlight (see Chapter 15) is array bounds checking. To simplify things I've taken the position that most presentations adopt and pretended that arithmetic overflow — breaking MININT, MAXINT limits — isn't a problem. I have, however, taken notice of division by zero, remaindering by zero, and array-bound errors. That enables me to illustrate the treatment of the problem without drowning in its details.

In informal proofs we usually treat definedness casually, not mentioning it unless it's a problem. Jape can't be casual, though it does try to deal with definedness in the background and hide the details from you whenever it can. It's quite good at that, but its powers are limited because the issue is quite subtle. To deduce $A < B + 1$ from $A \leq B$, for example, might seem irrefutable, but in the case that $A = B$ we need to know that both A and B are finite. In an informal proof we can look for evidence that A or B is finite or that $A \neq B$. But that's a bit much for a simple tool like Jape to handle automatically, so I didn't encode that particular deduction, or the others like it, and you have to rely on the 'obviously' step (see page 193) more than you might expect.

13.2 What the language means

Programs are specified by Hoare triples (Section 12.3). The meaning of the programming language is given in inference rules using Hoare triples: axioms for skip and assignment, rules with antecedents for everything else. Rules for loops and array-element assignment are given in later chapters.

Definition 13.1 $\{A\}\,\mathsf{skip}\,\{A\}$

skip does nothing, immediately and instantaneously, and always terminates. It has no effect at all (just like adding zero in arithmetic).

Definition 13.2
$$\frac{\begin{array}{cc}\vdots & \vdots \\ \{A\}\,\mathit{prog1}\,\{B\} & \{B\}\,\mathit{prog2}\,\{C\}\end{array}}{\{A\}\,\mathit{prog1}\,;\mathit{prog2}\,\{C\}}\ \mathit{sequence}$$

Semicolon-separated sequences have the effect of the first part of the sequence, then the effect of the second part. When we're being extremely formal, semicolon is left-binding — i.e. $A; B; C$ is read as $(A; B); C$ — but it's provably associative so it doesn't really matter.

Definition 13.3
$$\frac{\begin{array}{cc}\vdots & \vdots \\ \{A\}\,\mathit{prog1}\,\{B\} & \{B\}\,\mathit{prog2}\,\{C\}\end{array}}{\{A\}\,\mathit{prog1}\,\{B\}\,\mathit{prog2}\,\{C\}}\ \mathit{Ntuple}$$

Because it's sometimes convenient to define the intermediate formula B in a sequence, I allow a notational variant of the sequence rule in which the intermediate assertion is explicit in the conclusion.

13.3 Rules of consequence

Sometimes it is necessary to prove that a program is more generous or more precise than its specification. The need arises frequently, as examples will demonstrate. The **consequence** rules use implication to describe relaxation of a pre- or post-condition.

Definition 13.4
$$\frac{\begin{array}{cc}\vdots & \vdots \\ A \rightarrow B & \{B\}\,\mathit{prog}\,\{C\}\end{array}}{\{A\}\,\mathit{prog}\,\{C\}}\ \mathit{consequence}\ (L)$$

If a state which satisfies A must logically satisfy B, and a program started in B must reach C, then clearly a program started in A must reach C. The program is more capable than it needs to be: it works in states — inside the B-set of states but outside A — that the specification $\{A\}\,\mathit{prog}\,\{C\}$ doesn't require it to.

Definition 13.5
$$\frac{\begin{matrix}\vdots & & \vdots \\ \{A\}\,prog\,\{B\} & & B \rightarrow C\end{matrix}}{\{A\}\,prog\,\{C\}}\ consequence\ (R)$$

If a program started in a state which satisfies A must reach a state which satisfies B, and B logically guarantees C, then clearly a program started in A must reach C. The program is more precise than it needs to be: it always hits a particular part of the C-set target, even though the $\{A\}\,prog\,\{C\}$ specification would let it hit any part.

13.4 The marvellous definition of assignment

Hoare's definition of the effect of the assignment program is startling, marvellous, amazing, wonderful. It may be the most surprising thing a programmer ever learns. It captures effortlessly the way that an assignment changes a state. And it works in quite the opposite direction than you might have imagined: right-to-left from final state to starting state, even though execution goes the other way.

First a bit of notation: if A is a formula which may contain occurrences of the name x, then A^x_E is the same formula in which every occurrence of x has been replaced by the formula E.[3] For example, $(i > 5)^i_{i+1}$ is $i + 1 > 5$ which, as any fule kno,[4] is the same as $i > 4$.

Definition 13.6 $\{(E\ computes) \wedge A^x_E\}\ x := E\ \{A\}$

If you want an assignment to terminate in a state satisfying a postcondition A, you have to start it in a state which is ready for the assignment. We can describe that starting state using a copy of A in which claims about x have been replaced by claims about E — that is, A^x_E. After the assignment, we can refer to the assigned value using the name x; before the assignment, we have to call that value E; in either situation the same conditions must apply to it. Because the assignment affects only x, nothing else in the formula has to change.

The startling thing about Hoare's variable-assignment axiom is that it works backwards, right-to-left, although assignment executions work forwards, left-to-right. The beauty of it is that it replaces complicated questions about memory states with a simple formal calculation, an easy substitution of a formula for a name. The problem of talking about the pre-assignment value of x, which would

[3] Strictly, every **free** occurrence of x, but I shall be able to avoid that technical complication by using x, y, z as quantified variables and i, j, k as program variables, so I'm not going to explain it here.

[4] A Nigel Molesworth phrase, a perfect description of handwaving. See "Down with Skool" (Whillans and Searle) for more gems of 1950s UK prep-school comedy.

be necessary in an axiom that worked left-to-right, has been magicked away. Superb!

Actually there is one slight difficulty. The axiom needs a side-condition, a **no aliasing** condition on x, explained in Section 13.5 below. In variable assignment it isn't much of a problem, so I can defer it until you've seen some examples, but it will menace us again in Chapter 15.

13.4.1 You already know about substitution. You can think of the variable-assignment axiom as

$$\{(E \text{ computes}) \wedge P(E)\}\, x := E\, \{P(x)\}$$

— provided that you make sure there are no xs lurking in P. That is, the axiom is doing nothing you didn't already understand from the discussion of predicates and quantifiers in Chapter 6. But since what you actually have to do is cross out all the xs and replace them with Es, the A^x_E notation is clearer and more direct.

13.5 Some examples of assignment

13.5.1 Increase the value of a variable. You know, and I know, and every programmer knows that $i := i + 1$ is a program which increases by one the integer value stored in variable i. You know, and I know, and every programmer knows that if we want this program to finish in a state in which $i = 3$, then we have to start in a state in which $i = 2$, just because $2 + 1 = 3$. You know, and I know, and every programmer knows that no other value of i will do.

The assignment axiom confirms what we knew: $\{(i{=}3)^i_{i+1}\}\, i{:=}i{+}1\, \{i{=}3\}$, and $(i = 3)^i_{i+1}$ is $i + 1 = 3$ which, at the wave of a hand, simplifies to $i = 2$.

13.5.2 Verification conditions. Jape doesn't have hands to wave. If you set it the $\{i = 2\}\, i := i + 1\, \{i = 3\}$ problem and try to apply the variable-assignment axiom you get the result shown in Fig. 13.1. Jape has automatically and helpfully inserted a consequence(L) step because the specified precondition $i = 2$ isn't exactly the $i + 1 = 3$ that the axiom calculates. The result is the implication on line 1, an extra proof obligation that you might not have expected.

```
...
1: i=2→i+1=3
2: {i+1=3}(i:=i+1){i=3}   variable-assignment
3: {i=2}(i:=i+1){i=3}     consequence(L) 1,2
```

Fig. 13.1 Consequence and variable-assignment generate a verification condition

1:	i=2	assumption
2:	2+1=3	obviously
3:	i+1=3	equality-substitution 1,2
4:	i=2→i+1=3	→ intro 1-3
5:	{i+1=3}(i:=i+1){i=3}	variable-assignment
6:	{i=2}(i:=i+1){i=3}	consequence(L) 4,5

Fig. 13.2 Resolution of a verification condition

Implications generated automatically by use of the program logic rules are called **verification conditions**. Proving verification conditions is the real business of proving programs, because the program logic steps are mechanical and mindless. The business you have to do, as in this case, is more often arithmetical than logical even though it's expressed as a logical claim.

Jape doesn't do arithmetic,[5] but the Hoare-logic encoding can use **equality substitution** — replacement of equals by equals — to reduce line 1 to the point where it's **really** obvious: see lines 3 and 2 of Fig. 13.2.

Equality substitution as an inference rule is

$$\frac{\begin{matrix}\vdots & & \vdots \\ A = B & & P(B)\end{matrix}}{P(A)}$$

It's a feature of my encoding of Hoare logic for Jape, since so much of what has to be proved has to do with equality.

To close the proof on line 2 I used an 'obviously' step, a mechanism which I included in Jape's version of Hoare logic just to deal with arithmetic claims which are obviously true but difficult or impossible to establish with the rest of the encoding. An 'obviously' step will accept any conclusion and any hypotheses that you ask it to. Of course such a powerful proof cannon makes it easy to cheat, but of course I won't do that! (I do think I might have been justified in saying that line 1 of Fig. 13.1 was obvious, though.)

In this example there's no sign of the ($i + 1$ computes) definedness condition of the variable-assignment axiom. That's because it's dismissed behind the scenes: i is a variable, 1 is a constant and Jape ignores the possibility of overflow in addition.

[5] At the time of writing (September 2004) I haven't hitched Jape to any of the available 'arithmetic oracle' programs which it could use to solve this kind of problem mechanically. By the time you read this it might have been done, and then I wouldn't need to apologize. But I wouldn't apologize anyway: arithmetic isn't the game I want to play.

```
...
1: {i=Ki∧j=Kj}(t:=i){_B4}
...
2: {_B4}(i:=j){_B2}
...
3: {_B2}(j:=t){i=Kj∧j=Ki}
4: {i=Ki∧j=Kj}(t:=i;i:=j;j:=t){i=Kj∧j=Ki} sequence 1,2,3
-----------------------------------------------------------
Provided:
DISTINCT i, j, t
```

Fig. 13.3 'Unknown' pre- and post-conditions in a sequence of assignments

13.5.3 Take an increased value from another variable. What is the state in which we must start the program $i := j + 1$, if finally we must have $i = 3$? Surely it is $(i = 3)^i_{j+1}$, which is $j + 1 = 3$, which is $j = 2$. The precondition $j = 2$ doesn't mention i, because the initial value of i doesn't matter: it's going to be overwritten. The postcondition $i = 3$ doesn't mention j, because it doesn't matter where the value came from.

If we change the postcondition to $i = 3 \wedge j = 2$ then the variable-assignment axiom calculates the precondition $(i = 3 \wedge j = 2)^i_{j+1}$, which simplifies to $j + 1 = 3 \wedge j = 2$, and that's provably equivalent to $j = 2$. The longer postcondition could still be useful, because it emphasizes that the assignment changes only i.

If we make the postcondition $i = 3 \wedge j = 3$ the axiom calculates the precondition $(i = 3 \wedge j = 3)^i_{j+1}$, which is $j + 1 = 3 \wedge j = 3$, which implies $j = j + 1$, which is an arithmetical contradiction. That tells us that there is no state from which the program $i := j + 1$ will establish $i = 3 \wedge j = 3$.

13.5.4 Exchange the values of two variables. We'd like to be sure that the famous three-step variable-value-exchange program using an extra variable t really does work: i.e.

$$\{i = Ki \wedge j = Kj\}\, t := i;\ i := j;\ j := t\, \{i = Kj \wedge j = Ki\}$$

Here Ki and Kj are parameters of the problem describing the initial values of the variables; you can think of them either as constants or as formulae which don't mention i or j, so that substitution for i or for j has no effect on them. Notice that the pre- and postcondition say nothing about t: it doesn't matter what its initial value is, and we don't care about its final value — sometimes it's ok to hide in Hoare logic's small print and not to say everything that might be said.

Using the sequence rule, the assignment axiom and one of the rules of consequence we can work backwards from the desired final state. It's a simple,

...

1: {i=Ki∧j=Kj}(t:=i){_B4}

...

2: {_B4}(i:=j){i=Kj∧t=Ki}

3: {i=Kj∧t=Ki}(j:=t){i=Kj∧j=Ki} variable-assignment

4: {i=Ki∧j=Kj}(t:=i;i:=j;j:=t){i=Kj∧j=Ki} sequence 1,2,3

Provided:
DISTINCT i, j, t

Fig. 13.4 One intermediate formula calculated

mechanical and mindless calculation working backwards from the sequence and then backwards through the assignments, but the way it happens in Jape is quite revealing.

Fig. 13.3 shows the first step, reducing the sequence to its components and exposing the individual assignments. (The sequence rule deals with two-element sequences; for convenience I've made Jape deal with longer sequences in a single step.) The precondition of the sequence is the precondition of line 1, and the postcondition of the sequence is the postcondition of line 3. Inside the sequence the postcondition of line 1 is the precondition of line 2 and the postcondition of line 2 is the precondition of line 3. The intermediate formulae (B in the sequence rule) haven't been calculated yet, so Jape has used **unknowns**, in this case _B4 and _B2, to stand in for them. Unknowns look surprising at first, but they are no more than placeholders for formulae that you haven't yet decided on. They are easy to recognize: they are the only identifiers that start with an underscore.

The assignment axiom works backwards, so line 3 is the one to solve first (Jape will let you do the assignment steps in any order, but this is the easy way!). If you apply the variable-assignment axiom to that line Jape calculates what _B2 must be, replacing instances of j by t in the postcondition as the axiom requires and rewriting lines 2 and 3 to give Fig. 13.4 (it hides the 't computes' part of the precondition because variables always compute). Fig. 13.5 shows the result of applying the variable-assignment axiom again, this time to line 2. Finally, the same axiom plus consequence(L) gives Fig. 13.6. Proof of the verification condition is straightforward, using $\rightarrow$ intro, $\wedge$ intro and $\wedge$ elim, and not shown.

13.5.5 Anti-aliasing. In each of the Figs. 13.3, 13.3, 13.5 and 13.6 there's a **proviso** line — `DISTINCT i,j,t` — at the bottom of the proof. This is Jape's way of applying the variable-assignment axiom's **no aliasing** side condition, and it's time to explain what is going on.

```
...
1: {i=Ki∧j=Kj}(t:=i){j=Kj∧t=Ki}
2: {j=Kj∧t=Ki}(i:=j){i=Kj∧t=Ki}           variable-assignment
3: {i=Kj∧t=Ki}(j:=t){i=Kj∧j=Ki}           variable-assignment
4: {i=Ki∧j=Kj}(t:=i;i:=j;j:=t){i=Kj∧j=Ki} sequence 1,2,3
------------------------------------------------------------
Provided:
DISTINCT i, j, t
```

Fig. 13.5 Both intermediate formulae calculated

```
...
1: i=Ki∧j=Kj→j=Kj∧i=Ki
2: {j=Kj∧i=Ki}(t:=i){j=Kj∧t=Ki}           variable-assignment
3: {i=Ki∧j=Kj}(t:=i){j=Kj∧t=Ki}           consequence(L) 1,2
4: {j=Kj∧t=Ki}(i:=j){i=Kj∧t=Ki}           variable-assignment
5: {i=Kj∧t=Ki}(j:=t){i=Kj∧j=Ki}           variable-assignment
6: {i=Ki∧j=Kj}(t:=i;i:=j;j:=t){i=Kj∧j=Ki} sequence 3,4,5
------------------------------------------------------------
Provided:
DISTINCT i, j, t
```

Fig. 13.6 Verification condition emerges

Hoare's assignment axiom works only if there is no other name for the program variable called x. If x and y name the same variable, for example, then necessarily $x = y$ and therefore necessarily $x \neq y + 1$. But Definition 13.6 seems to say that

$$\{(x = y + 1)^{x}_{y+1}\}\, x := y + 1\, \{x = y + 1\}$$

and, as any fule will tell you, $(x = y + 1)^{x}_{y+1}$ simplifies to $y + 1 = y + 1$, which is $\top$. So the assignment axiom seems to tell us that

$$\{\top\}\, x := y + 1\, \{x = y + 1\}$$

We know that $\top$ holds in any state, and because x and y name the same variable we know that $x = y+1$ can't hold in any state. It seems that with the assignment axiom we can establish $\perp$ any time we like. That's a paradox!

Aliasing in programs, where the same variable has more than one name, happens more often than you might expect: it's a real problem with array elements, as we'll see in Chapter 15; it occurs all the time in object-oriented programs (if i and j are references to the same object, then $i.x$ and $j.x$ are aliases); in some languages, notably Pascal and C++, different procedure parameters can name the same argument-variable.

The problem isn't really with the assignment axiom, it's with substitution, and the solution is to be more careful when replacing a variable name with a formula. We ought to say that A^x_E means "replace each occurrence of x or an alias of x with a copy of E". So if x and y are aliases, as in the example above, $(x = y + 1)^x_{y+1}$ should be $y + 1 = (y + 1) + 1$, which is $y + 1 = y + 2$, which is $\perp$; the precondition is as impossible as the postcondition, and the paradox has gone.

In practice, though, alias-sensitive substitution is too difficult to use in informal paper or blackboard proofs. So the assignment axiom has a side-condition: there must be **no aliases** for x in A. Then we can read A^x_E as "cross out the xs and insert Es instead", and it all works properly.

Most Hoare logic presentations don't make a fuss about aliasing, and in informal proofs we usually skate over the problem. Jape has to be more careful. In Fig. 13.6, so far as Jape is concerned, we are proving a general theorem about a program in which the variable names i, j and t are parameters of the proof, and only those instances of the theorem in which the three variables are distinct are valid. If you break that distinctness condition you don't get a valid result: $k := i$; $i := k$; $k := k$, for example, in which I've put k in place of both j and t, and thus broken the distinctness condition, doesn't do a variable exchange.

Jape is built to do substitution very carefully. Without the distinctness proviso it couldn't simplify the substitutions A^x_E generated by the variable-assignment axiom. Fig. 13.7 shows the effect of trying a variable-assignment step without a proviso to help. Jape can't be sure on line 3 whether i^j_t should be i (i and j distinct) or t (i and j aliases), so it leaves the matter undecided. It looks horrid, it is horrid, and it gets worse if you carry on, because the conjecture simply isn't valid without the proviso, and therefore, by soundness, can't be proved. That's why I use distinctness provisos in my examples.

13.5.6 A touch of informality. The proof in Fig. 13.6 is repetitive: there are twelve instances of only five formulae in pre and postconditions and in the verification condition. On paper and on the blackboard we try to cut down that

```
   ...
1: {i=Ki∧j=Kj}(t:=i){_B4}
   ...
2: {_B4}(i:=j){i«t/j»=Kj∧t=Ki}
3: {i«t/j»=Kj∧t=Ki}(j:=t){i=Kj∧j=Ki}         variable-assignment
4: {i=Ki∧j=Kj}(t:=i;i:=j;j:=t){i=Kj∧j=Ki}   sequence 1,2,3
```

Fig. 13.7 The signs of aliasing

repetition. We deal with sequences of assignments as a matter of calculation, and hide the proof structure entirely if we can. That is, we work backwards to see what pre-condition A' is required by considering only the post-condition B in $\{A\}\,prog\,\{B\}$, and then we see if $A \rightarrow A'$. We do the same calculations as in Fig. 13.6, but we don't mention the proof structure because it's straightforward and absolutely standard.

The calculation so far as variable assignment is concerned can be laid out vertically as an Ntuple, where each step is an application of the variable-assignment axiom:

$$
\begin{array}{l}
\{j = Kj \wedge i = Ki\} \\
\quad t := i \\
\{j = Kj \wedge t = Ki\} \\
\quad i := j \\
\{i = Kj \wedge t = Ki\} \\
\quad j := t \\
\{i = Kj \wedge j = Ki\}
\end{array}
$$

The structure that's being hidden here is the sequence rule, which glues together the intermediate specifications.

The consequence step can be added at the head of the proof, if it needs to be included — but we don't make more fuss about it than we absolutely must.

$$
\begin{array}{l}
\{i = Ki \wedge j = Kj\} \;\therefore \\
\{j = Kj \wedge i = Ki\} \\
\quad t := i \\
\{j = Kj \wedge t = Ki\} \\
\quad i := j \\
\{i = Kj \wedge t = Ki\} \\
\quad j := t \\
\{i = Kj \wedge j = Ki\}
\end{array}
\tag{13.1}
$$

Calculation (13.1) conveys to an expert the same information as the proof in Fig. 13.6, but far more succinctly. It can stand in for the proof provided that you realize it's hiding uses of the sequence rule, and waving its hands over the implication in the consequence step. That's usually ok in practice, because we are more interested in the calculations than the rules.

In this book I've taken the position that you can learn to be informal once you realize what's going on formally. (13.1) isn't completely formal, because it doesn't say what rules it's using and it hides some steps, but it's what I'd write on paper or on a blackboard if asked. I've made Jape show the formal proof structure because that is what I want you to understand before you move on to informality.

13.5.7 A non-example. Novices often get the variable-exchange program wrong. Not realizing the destructive power of assignment, they hope that $i := j;\ j := i$ will do the job. Of course it won't — at least, not often:

$$
\begin{array}{l}
\{i = Ki \land j = Kj\} \ \therefore \ \ ?? \\
\{j = Kj \land j = Ki\} \\
\quad i := j \\
\{i = Kj \land i = Ki\} \\
\quad j := i \\
\{i = Kj \land j = Ki\}
\end{array}
$$

To prove that $i = Ki \land j = Kj \rightarrow j = Kj \land j = Ki$ you need $Ki = Kj$ or $i = j$. So, like a broken watch which shows the time correctly twice a day, the non-exchange program works, but only when there's no need for it.

13.6 The definition of choice

The variable-assignment axiom makes it possible to calculate a precondition from a stated postcondition. I've chosen a definition of choice which has the same property. It's reminiscent of the $\lor$ elim rule of Chapter 3.

Definition 13.7

$$
\frac{\begin{array}{cc} \vdots & \vdots \\ \{A\}\ \mathit{prog1}\ \{C\} & \{B\}\ \mathit{prog2}\ \{C\} \end{array}}{\{(E\ \mathit{computes}) \land (E \rightarrow A) \land (\neg E \rightarrow B)\}\ \textsf{if}\ E\ \textsf{then}\ \mathit{prog1}\ \textsf{else}\ \mathit{prog2}\ \textsf{fi}\ \{C\}}\ \mathit{choice}
$$

A choice program executes *prog1* when E evaluates to $\top$, *prog2* when E evaluates to $\bot$ (and, of course, $\neg\top \equiv \bot$ and $\neg\bot \equiv \top$). The precondition which guarantees that a choice will reach a state C is that E can be safely evaluated, that when E holds so does the precondition for *prog1* to reach C, and that on the other hand when $\neg E$ holds so does the precondition for *prog2* to reach just the same state C.

13.7 An example with choice

Assignment proofs, even of sequences of assignments, are pretty straightforward. Choice proofs are a bit more complicated, and since negation is part of the choice rule they usually involve contradiction steps.

Realizing that many of my readers will be university students, I've chosen an example which may be close to their hearts, a program which associates a pass/fail grade with a numerical mark. In many UK universities, 40 is the pass

```
...
1: i=Ki→(i≥40→_A5)∧(¬(i≥40)→_B6)
...
2: {_A5}(r:=pass){i=Ki∧(i<40→r=fail)∧(i≥40→r=pass)}
...
3: {_B6}(r:=fail){i=Ki∧(i<40→r=fail)∧(i≥40→r=pass)}
4: {(i≥40→_A5)∧(¬(i≥40)→_B6)}if i≥40 then r:=pass else r:=fail fi    choice 2,3
   {i=Ki∧(i<40→r=fail)∧(i≥40→r=pass)}
5: {i=Ki}if i≥40 then r:=pass else r:=fail fi                          consequence(L) 1,4
   {i=Ki∧(i<40→r=fail)∧(i≥40→r=pass)}
```

Provided:
DISTINCT i, r

Fig. 13.8 The choice rule appears to make a mess

mark and 70 or 80 is the best that a genius can hope for; 80 is very much more than twice as good as 40, of course.[6]

$$\begin{array}{l}\{i = Ki\} \\ \quad \text{if } i \geq 40 \text{ then } r := pass \text{ else } r := fail \text{ fi} \\ \{i = Ki \wedge (i < 40 \rightarrow r = fail) \wedge (i \geq 40 \rightarrow r = pass)\}\end{array} \tag{13.2}$$

The precondition states a starting value for i. The postcondition states that the value of i doesn't change, and gives conditional formulae to describe what the final value of r must be. It looks like the natural specification, and it looks very straightforward.

Fig. 13.8 shows the effect of applying the choice rule to this problem in Jape. Just as in Fig. 13.3, Jape's used unknowns (in this case _A5 and _B6) in place of the intermediate formulae A and B of the rule. Just as in several previous examples it's generated a verification condition using consequence(L) because the specified precondition $i = Ki$ looks nothing like the precondition the rule generates. Just as before, the unknowns are resolved by use of the assignment axiom, and the result is Fig. 13.9.

The verification condition on line 1 looks absolutely ferocious. Laying it out a bit better than Jape can manage, it reads

$$i = Ki \rightarrow \left(\begin{array}{l} \left(i \geq 40 \rightarrow i = Ki \wedge \left(\begin{array}{l} (i < 40 \rightarrow pass = fail) \wedge \\ (i \geq 40 \rightarrow pass = pass) \end{array} \right) \right) \wedge \\ \left(\neg(i \geq 40) \rightarrow i = Ki \wedge \left(\begin{array}{l} (i < 40 \rightarrow fail = fail) \wedge \\ (i \geq 40 \rightarrow fail = pass) \end{array} \right) \right) \end{array} \right) \tag{13.3}$$

[6] That's not the only daft marking idea. What's the average of A and D?

...

1: i=Ki→(i≥40→i=Ki∧(i<40→pass=fail)∧(i≥40→pass=pass))
∧(¬(i≥40)→i=Ki∧(i<40→fail=fail)∧(i≥40→fail=pass))

2: {i=Ki∧(i<40→pass=fail)∧(i≥40→pass=pass)}
(r:=pass){i=Ki∧(i<40→r=fail)∧(i≥40→r=pass)} — variable-assignment

3: {i=Ki∧(i<40→fail=fail)∧(i≥40→fail=pass)}
(r:=fail){i=Ki∧(i<40→r=fail)∧(i≥40→r=pass)} — variable-assignment

4: {(i≥40→i=Ki∧(i<40→pass=fail)∧(i≥40→pass=pass))∧(¬(i≥
40)→i=Ki∧(i<40→fail=fail)∧(i≥40→fail=pass))}if i≥40 then
r:=pass else r:=fail fi{i=Ki∧(i<40→r=fail)∧(i≥40→r=pass)} — choice 2,3

5: {i=Ki}if i≥40 then r:=pass else r:=fail fi
{i=Ki∧(i<40→r=fail)∧(i≥40→r=pass)} — consequence(L) 1,4

Provided:
DISTINCT i, r

Fig. 13.9 The mess resolved by the variable-assignment axiom

1:	i=Ki	assumption
2:	i≥40	assumption
	...	
3:	i<40→pass=fail	
	...	
4:	i≥40→pass=pass	
5:	i=Ki∧(i<40→pass=fail)∧(i≥40→pass=pass)	∧ intro 1,3,4
6:	i≥40→i=Ki∧(i<40→pass=fail)∧(i≥40→pass=pass)	→ intro 2-5
	...	
7:	¬(i≥40)→i=Ki∧(i<40→fail=fail)∧(i≥40→fail=pass)	
8:	(i≥40→i=Ki∧(i<40→pass=fail)∧(i≥40→pass=pass)) ∧(¬(i≥40)→i=Ki∧(i<40→fail=fail)∧(i≥40→fail=pass))	∧ intro 6,7
9:	i=Ki→(i≥40→i=Ki∧(i<40→pass=fail)∧(i≥40→pass=pass)) ∧(¬(i≥40)→i=Ki∧(i<40→fail=fail)∧(i≥40→fail=pass))	→ intro 1-8

Fig. 13.10 A partly dismantled verification condition

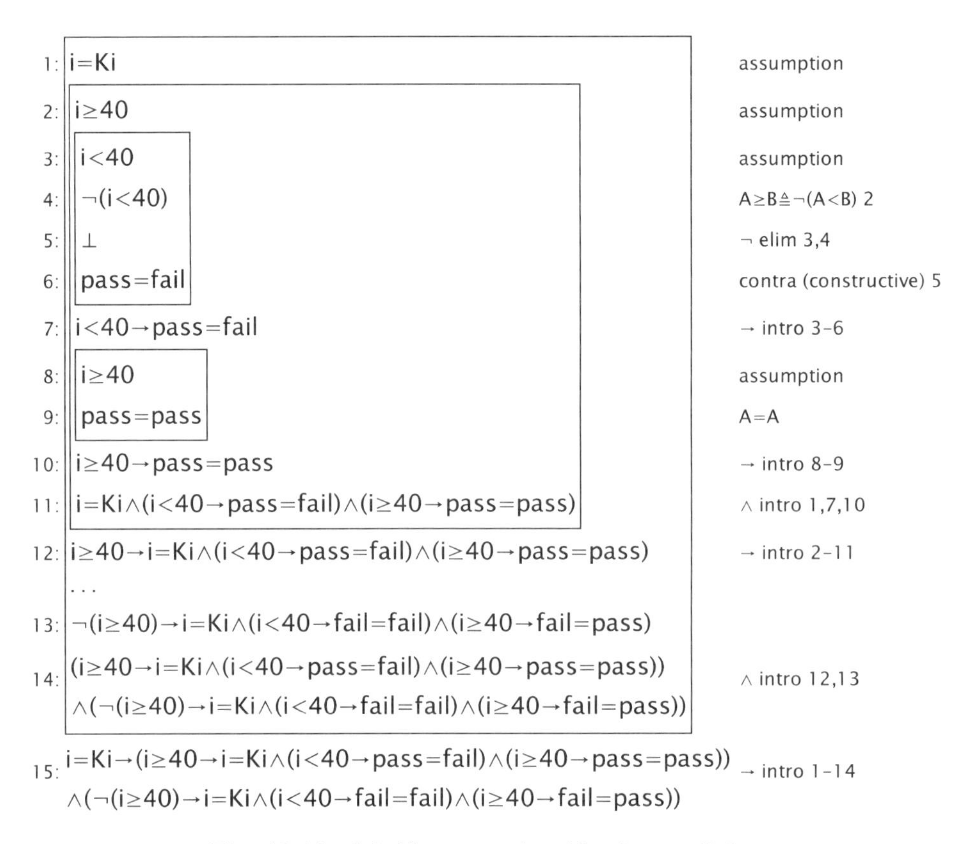

Fig. 13.11 A half-conquered verification condition

This formula looks completely stupid — it has $pass = fail$ and $fail = pass$ inside it, for example — and it's awfully long. But it is what the rule calculates, so I press on.

Making a lemma of the verification condition and taking the obvious backwards steps in Jape I arrive at Fig. 13.10. The problem isn't so hard after all! The implication on line 4 is trivial price-of-tomatoes stuff, and the one on line 3 is, of course, cunning-uncle business because $i \geq 40 \wedge i < 40$ is a contradiction. That much of the proof is shown in Fig. 13.11; the other half proceeds very similarly.

It's pleasing, I think, to see that those at first peculiar and now familiar rules of implication and negation mesh so well with calculation, so that it's possible to define real programming problems in logical notation and work them through to a conclusion.

14 Loops

Most interesting programs use loops (or recursion, which is just a super kind of loop). Hoare logic's treatment of loops is by the method of **invariants**, and it takes some swallowing. Once swallowed, though, never forgotten ... and it definitely is worth the effort.

14.1 How to specify the effect of a loop

To begin, look at a picture of the execution of while E do *prog* od, drawn as a flowchart in Fig. 14.1. Action follows the arrowed lines, starting at the top and exiting, if the loop ever terminates, at middle right. The diamond indicates that the formula E is evaluated; the labelled lines show that if E fails, action follows the horizontal line and the loop terminates; if E holds, action follows the vertical line. This shows that when a loop terminates, we can be sure that $\neg E$ is satisfied.

The box indicates execution of *prog*. Because execution of the box follows a successful test, whenever *prog* is executed E must hold. The line leading out of the *prog*-box shows that after *prog* is executed, action moves to evaluation of E again. It's a straightforward picture of while E do *prog* od, repeatedly testing E and executing *prog* until eventually the test fails.

14.1.1 A loop with a precondition. Suppose we start a loop in a state which satisfies a formula A. Fig. 14.2 shows what must happen on the first execution. If the E-test fails, the loop exits immediately — and then we know that the state satisfies $A \wedge \neg E$. If the E-test succeeds, then *prog* is executed, in a state which satisfies $A \wedge E$. After *prog* finishes, the state will satisfy some formula or other — call it B. Then the next execution would follow the same analysis, but starting with B rather than A, and ending up with C, ... and so on.

That is all very well, but it doesn't cut the mustard. By using A, then B, then C, and so on, it makes reasoning about each execution of the loop different from every other. They are indeed different executions, but if we have to treat them differently we will never get anywhere: reasoning has to overcome the loop problem, not reproduce it. It doesn't deal with the fact that reasoning about *prog* often has to go backwards from B to $A \wedge E$ — which means that we would need to know B to start with. And the labelling of the flowchart is wrong: two lines join at the top, one labelled A, the other B. It's all a bit of a mess.

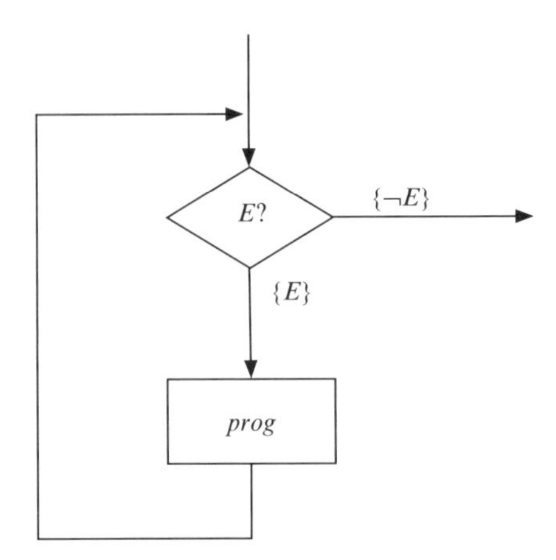

Fig. 14.1 Flowchart of while E do *prog* od

14.1.2 Invariant formulae. There's only one known way to solve these problems. Surprisingly, we make A and B the same formula! Calling them both I for **invariant**, the flowchart looks like Fig. 14.3. Now every execution of the loop is the same as every other, so far as formulae are concerned. Each starts in a state described by I and either executes *prog* in a state described by $I \wedge E$, or exits in a state described by $I \wedge \neg E$. It all works perfectly, provided that $\{I \wedge E\}\, prog\, \{I\}$.

14.1.3 No, it isn't a stupid idea. The invariant I is a **formula**, not a value. A specification formula can depend on program variables, and an invariant will

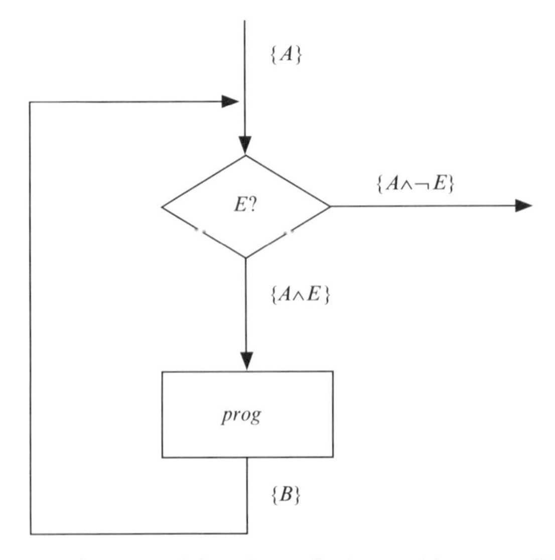

Fig. 14.2 Annotated flowchart of a loop with precondition A

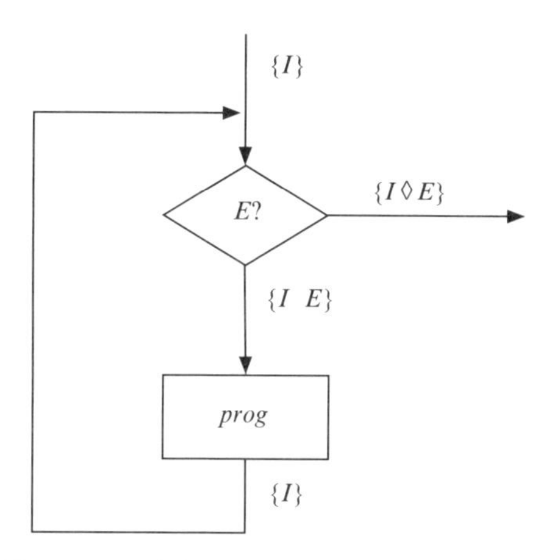

Fig. 14.3 A loop with invariant formula I

necessarily depend on them. If the values of the variables change, you would expect the value of the formula to change too. But what if I expresses a relationship between the values of program variables, rather than a precise description? Then there can be very many states in which make the invariant hold, and the job of the loop is to steer a path through those states to a point where the values of the variables make the E-test fail, and the loop has gone far enough.

But it's still a surprising idea. Its best defence is that it works. Time for some examples!

14.2 An aside on integer division and remaindering

The examples I can show you at this stage depend on arithmetic. The ones I've chosen involve integer division and remaindering, so I have to remind you of stuff you may have forgotten, from the time you first met division and long before you encountered fractions.

In our first school we deal with problems like

> "Johnny has nineteen apples. He shares them between his three horses, a donkey and a cow. How many does each of them get? How many are left over?"

The answer, of course, is three each and four left over, provided you don't let the animals get too close to the left-over pile. Integer division is an operation which delivers two results: a **quotient** $A \div B$ — in Johnny's problem, the quotient $19 \div 5$ is 3 — and a **remainder** $A \bmod B$ — for Johnny, the remainder $19 \bmod 5$ is 4.

When later we learn about fractional arithmetic it's easy to jump to the conclusion that integer division is a fairy-tale told to children, and fractions are the real thing. For one thing, fractional multiplication and division cancel — $(19/5) \times 5 = 19$ — whereas the integer versions don't seem to — $(19 \div 5) \times 5 = 15$.

Fractional arithmetic seems to be more accurate, to follow rules more carefully, to be more formal than what we were first taught.

Of course that's a misunderstanding. Integer multiplication and division do cancel perfectly, provided you don't forget the remainder. $(A \div B) \times B + A \bmod B = A$, every time (provided, of course, that B isn't zero — but then fractional arithmetic has the same problem). Integer division is nicely defined even for negative numbers (by convention $0 \leq A \bmod B < |B|$ — i.e. the remainder is never negative).

One use of remaindering is to test when one integer is a multiple of another: $A \bmod B = 0$ just when A is a multiple of B, i.e. when B exactly divides A. Another use is to select part of a number: $A \bmod 10$, for example, is the last digit of the decimal representation of A, and $A \div 10$ is the value of the numeral to the left of the last digit, or zero if there's nothing there — $456 \bmod 10 = 6$ and $456 \div 10 = 45$, for example.

In computing, working with binary numerals, we often deal with $\div 2$ (same as a single-bit right shift) and mod 2 (same as selecting the last bit). It's essential in those circumstances to recall what integer division really means and not to pretend that we're working with fractions.

14.3 A loop example

Consider the problem of finding if an integer n is prime or not — that is, whether it has any factors other than itself and 1. The algorithm I used when I was a boy searched for a factor, and went like this:

- does n divide exactly by 2?
- if so, then I've found a factor and n isn't prime;
- if not, does n divide by 3?
- if so, then I've found a factor and n isn't prime;
- if not, does n divide by 4?
- ...

... and so on, until I had either found a factor or reached n itself, at which point I could conclude that n had no factors other than itself and 1 and was therefore prime. Later on I realized that I need only consider prime divisors (i.e. I could ignore composites 4, 6, 8, 9, 10, 12 and so on); later still it was pointed out to me that I could stop at $\sqrt{n}$ rather than n. I'm going to ignore those sophisticated enhancements, and investigate the naive technique.

Fig. 14.4 shows a formal version of the algorithm, which sets the variable *prime* to $\top$ or $\bot$ according to whether or not n is a prime integer. It works provided that $n \geq 2$: the invariant of the loop is that i is somewhere in the range

$i := 2;$
while $n \bmod i \neq 0$ do $i := i + 1$ od;
$prime := (i = n)$

Fig. 14.4 A naive primacy checker

2..n, that we have tested every number from 2 up to but not including i, and we've found that none of them is a factor of n.

$$I : 2 \leq i \leq n \wedge \forall x(2 \leq x < i \rightarrow n \bmod x \neq 0) \tag{14.1}$$

Before I show the rule which defines the meaning of loops, I'm going to illustrate its use by giving an informal argument that this program does what we think it should. Then I shall be in a position to show you the rule, and give a more formal proof.

> The program in Fig. 14.4 is concise, and doesn't include any tests to ensure that it's being used safely. The loop guard, for example, doesn't include a test that $i \leq n$, so the program will loop forever if $n < 2$. That's a characteristic of programs which are formally verified: you can afford to program dangerously if you can prove you have made no mistakes and you can require that the program will only be used carefully.

14.3.1 Step 1: the invariant holds before the loop starts. The program is only expected to do its work when $n \geq 2$. After the first assignment $i = 2$, and those two facts together give us the first part of the invariant: $2 \leq i \leq n$. The quantification in the invariant then contains a vacuous implication, because there are no xs such that $2 \leq x < 2$. So we have I on entry to the loop, as Fig. 14.3 requires.

14.3.2 Step 2: the loop guard doesn't crash. We can't ignore definedness when analysing this program. The calculation $n \bmod i \neq 0$ will crash if $i = 0$, and if that crashes the loop will crash and take the program with it. But according to Fig. 14.3 we can be sure of two things:

- when we reach the loop, the invariant I holds;
- each time round the loop, when we test the loop guard, I still holds.

So the loop won't crash if when I holds, $n \bmod i \neq 0$ is defined: that is, because comparison with 0 isn't dangerous, if $I \rightarrow$ ($n \bmod i$ computes). And that is so, because I guarantees $i \geq 2$, which means $i \neq 0$, which means $n \bmod i$ won't crash.

14.3.3 Step 3: the invariant is preserved by the loop. Just before the loop test we know I; inside the loop, after a successful test, we know $I \wedge n \bmod i \neq 0$. I tells us that no number in the range $2..i-1$ divides n; $n \bmod i \neq 0$ tells us that i doesn't divide n either. So we know that no number in the range $2..i$ divides n, and it's safe to increase i by 1.

It's also possible, though a little more difficult, to argue that $i+1$ won't exceed n. Here goes: $n \bmod i \neq 0 \rightarrow n \neq i$ (by contradiction because if $n = i$ then $n \bmod i = 0$). From I we know that $i \leq n$; from $n \bmod i \neq 0$ we realize that $i \neq n$; therefore $i < n$; therefore $i + 1 \leq n$. It **is** safe to increase i by 1.

14.3.4 Will it ever stop? Step 3 shows that $\{I \wedge E\}\, i := i{+}1\, \{I\}$. From Fig. 14.3 we can see that **if** the loop exits, we are guaranteed $I \wedge \neg E$. We seem to have all that we need to conclude that the program works. But we already know from McCarthy's millionaire joke (Fig. 12.3, page 184) that we haven't done enough. Every programmer knows that some innocent-looking loops buzz round for ever. We still have to prove that the loop really **does** exit.

Some loops are bound to exit given the right circumstances. Here's an algorithm for descending a finite staircase:

```
while ¬(on the ground floor) do
   take one step down
od
```

The algorithm doesn't terminate if you start in the basement (because you'd be walking in the wrong direction); it doesn't terminate if there is a gap in the staircase (you'd fall through); it doesn't terminate if the staircase is infinite (it would take too long to come down); but otherwise — start at or above the ground floor on a finite staircase with no holes in it — it's guaranteed to finish at the ground floor. That's because each execution subtracts one from the finite non-negative number of steps to be descended, and you can't do that for ever without reaching zero.

The principle of counting downwards to zero lies behind the Hoare-logic treatment of the termination of loops (and behind the closely related principle of mathematical induction). We need an estimate, called a variant or a **measure**, of the number of times a loop will execute, expressed as an integer formula M. The estimate must never be optimistic (too small), but it can be as pessimistic as may be (as large as you like). We can even take great jumps provided that we don't let the measure go negative: that is, provided we never take a step down the basement staircase.

If we can show that each execution of the loop body reduces the measure, our loop is safely counting down; if we can show that it never executes the body when the measure is negative, it won't miss the target and count down for ever.

14.3.5 Step 4: find a measure. To show that a loop will exit we pick a formula M and prove two things about it:

1. $I \wedge E \to M > 0$ (if we are about to make a step, we haven't already missed the ground floor);
2. $\{I \wedge E \wedge M = m\}\, prog\, \{M < m\}$ (we count downwards on every step).

(Here M, I and E are formulae whereas m is an arbitrary fixed constant, like the i of the $\forall$ intro and $\exists$ elim rules.)

In the naive prime-finding algorithm of Fig. 14.4 a measure of the number of executions still left to do is $n-i$. The loop may finish early by finding a factor, in which case it won't have to do as many as $n - i$ tests, or it may go on all the way to discover that n is prime, in which case it will have to reduce $n - i$ to zero, and then is certain to stop because $n \bmod n{=}0$.

14.3.6 Step 5: we don't go past the exit. Each time we are about to take a step, we have $I \wedge E$. To show we never start walking down to the basement, we have only to show $I \wedge E \to n - i > 0$. $I \wedge E$ tells us that $i < n$ (see step 3); therefore $0 < n - i$; therefore $n - i > 0$.

14.3.7 Step 6: we always move downwards. It's obvious that the program $i := i + 1$ reduces $n - i$, because it increases i.

14.3.8 Step 7: we have the postcondition we need. The while loop will terminate, according to Fig. 14.3, with $I \wedge \neg E$. We want to know that we have the precondition for the assignment $prime := n = i$ to tell us whether we have a prime number or not.

A number is prime if and only if it has no positive factors other than itself and 1. The postcondition of the program is

$$prime \iff \forall x(2 \leq x < n \to n \bmod x \neq 0)$$

— that is, *prime* should be set to $\top$ if and only if n is actually a prime number.

The assignment axiom, using the final assignment $prime := n = i$, transforms this postcondition into $n = i \iff \forall x(2 \leq x < n \to n \bmod x \neq 0)$. The consequence(L) rule tells us that we must prove

$$I \wedge n \bmod i = 0 \to \left(n = i \iff \forall x \left(2 \leq x < n \to n \bmod x \neq 0 \right) \right)$$

The hand-waving gets a little furious here. To prove logical equivalence ($\iff$) we have to prove an implication in either direction. It's obvious that when we have the quantification $\forall x(2 \leq x < i \to n \bmod x \neq 0)$ from the invariant and we have $n = i$ from the left-hand side of the equivalence then we have the quantification on the right (substitute n for i).

Going in the other direction is trickier, and needs an argument by contradiction. If i is in the range $2..n$, as the invariant says it is, and if $n \bmod i = 0$, and we have $\forall x(2 \leq x < n \rightarrow n \bmod x \neq 0)$, and $n \neq i$, then there's a contradiction — i is in the range of the right-hand quantification, and so we should have $n \bmod i \neq 0$. Reasoning classically, we must have $n = i$, and the logical equivalence is proved.

14.4 The while rule

Definition 14.1

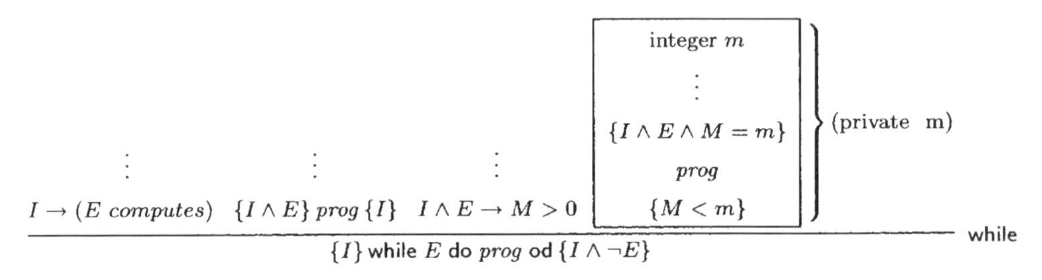

Provided that

- we start in a state satisfying invariant I; **and**
- I guarantees that the test formula E can be safely calculated; **and**
- I and E together guarantee that the loop body preserves the invariant; **and**
- I and E together guarantee that the measure M is strictly positive; **and**
- I and E together guarantee that the loop body reduces the measure: **then**
- we can be sure that the loop will terminate in a state satisfying $I \wedge \neg E$.

That doesn't sound much, but it's actually a great deal. It's the best handle on loops that there is. It tells us that the loop will terminate, and it describes the state it will terminate in.

To use the while rule you have to invent an invariant formula I and a measure formula M; they don't automatically emerge from the program itself. It isn't always easy to discover the right formulae, and the proofs aren't always easy even when you have discovered them. But loops are hard to write, go wrong often, and are the major cause of difficult programming bugs, so you should expect that to prove that a loop is built correctly will require intelligence, sharp tools, and a good deal of heavy lifting.

To prove $\{A\}$ while E do *prog* od $\{B\}$ for arbitrary A and B — i.e. when A isn't I and/or B isn't $I \wedge \neg E$ — we use the consequence rules, and prove separately $A \rightarrow I$ and $I \wedge \neg E \rightarrow B$.

14.5 A formal treatment of the primacy algorithm

The primacy algorithm isn't **that** interesting, but it is interesting enough to make a good formal proof example. The proof is in several parts, each of them quite intricate.

You may wonder why it's necessary to make a formal proof at all given that we already have a convincing informal argument. One answer is that formal proofs keep you honest: it's easy to make a slip in an informal argument whilst gliding over what looks like irrelevant details. But the real answer is that making formal proofs gives you experience that you can use to make and understand informal arguments.

> I repeat the warning given in the introduction to this part of the book: the proofs which follow are quite large and the proof searches are not given in detail. It may be best to reproduce them in Jape for yourself as you follow the discussion.

14.5.1 Verification conditions. To save space on the page I've separated off the last line of the program, the assignment to *prime*, leaving the initialization and the while loop. The first two steps in Jape (sequence, while) produce the effect shown in Fig. 14.5. Almost all the Hoare-logic work is already done: we only have to generate verification conditions from the assignments on lines 3 and 6, use Jape's Unify command to tell it that `_M` should be $n - i$, and that's it. The rest of the work — and it's most of the work — is a formal slog through the verification conditions.

There's a consequence step (line 9) in the proof already. This time it's consequence(R) because the postcondition on the while (line 7) is not exactly the same as the one we need to prove on line 10. There are two verification conditions already on lines 2 and 4, and the assignments on lines 1, 3 and 6 will generate three more. The verification condition on line 2 is the remains of the definedness condition $I \rightarrow (E \text{ computes})$.

The box on lines 5–6 comes from the right-hand antecedent of the rule. It's there to remind us that Km in the proof (m in the rule) is really private to that sub-proof, to prevent cheating when arguing that the measure decreases.

As usual, there's an anti-aliasing proviso at the bottom of the proof. There are only two variables in the program (Km is both private and a constant, so can be ignored on two counts), but they still have to be distinct.

The presence marker on line 5 is 'integer Km' where you might have expected 'actual Km'. Hoare logic quantifies only over integers, and I've tweaked the treatment of presence and quantification to match: you'll see that in several of the proofs.

```
   ...
1: {n≥2}(i:=2){2≤i∧i≤n∧∀x.(2≤x∧x<i→n mod x≠0)}
   ...
2: 2≤i∧i≤n∧∀x.(2≤x∧x<i→n mod x≠0)→i≠0
   ...
3: {2≤i∧i≤n∧∀x.(2≤x∧x<i→n mod x≠0)∧n mod i≠0}
   (i:=i+1){2≤i∧i≤n∧∀x.(2≤x∧x<i→n mod x≠0)}
   ...
4: 2≤i∧i≤n∧∀x.(2≤x∧x<i→n mod x≠0)∧n mod i≠0→_M>0
5: | integer Km                                            assumption
   | ...
6: | {2≤i∧i≤n∧∀x.(2≤x∧x<i→n mod x≠0)
   | ∧n mod i≠0∧_M=Km}(i:=i+1){_M<Km}
   {2≤i∧i≤n∧∀x.(2≤x∧x<i→n mod x≠0)}
7: while n mod i≠0 do i:=i+1 od                           while 2,3,4,5-6
   {2≤i∧i≤n∧∀x.(2≤x∧x<i→n mod x≠0)∧¬(n mod i≠0)}
   ...
8: 2≤i∧i≤n∧∀x.(2≤x∧x<i→n mod x≠0)∧¬(n mod i≠0)
   →2≤i∧i≤n∧∀x.(2≤x∧x<i→n mod x≠0)∧n mod i=0
   {2≤i∧i≤n∧∀x.(2≤x∧x<i→n mod x≠0)}
9: while n mod i≠0 do i:=i+1 od                           consequence(R) 7,8
   {2≤i∧i≤n∧∀x.(2≤x∧x<i→n mod x≠0)∧n mod i=0}
   {n≥2}(i:=2){2≤i∧i≤n∧∀x.(2≤x∧x<i→n mod x≠0)}
10: while n mod i≠0 do i:=i+1 od                          Ntuple 1,9
   {2≤i∧i≤n∧∀x.(2≤x∧x<i→n mod x≠0)∧n mod i=0}
-----------------------------------------------------------------------
Provided:
DISTINCT i, n
```

Fig. 14.5 The beginning of a proof of the primacy checker

14.5.2 Establishing the invariant. Line 1 of Fig. 14.5 states that the assignment $i := 2$, given precondition $n \geq 2$, establishes the invariant we need for the loop on line 7. The standard phrase is "the initialization establishes the invariant". A proof of the verification condition it generates is shown in Fig. 14.6.

The formal proof follows the same track as the informal argument in Section 14.3.1. The basic logical rules bring it home. You don't need to know more about implication and quantification than the introduction and elimination rules of Chapter 3 and Chapter 6, tweaked slightly to recognize that we quantify only

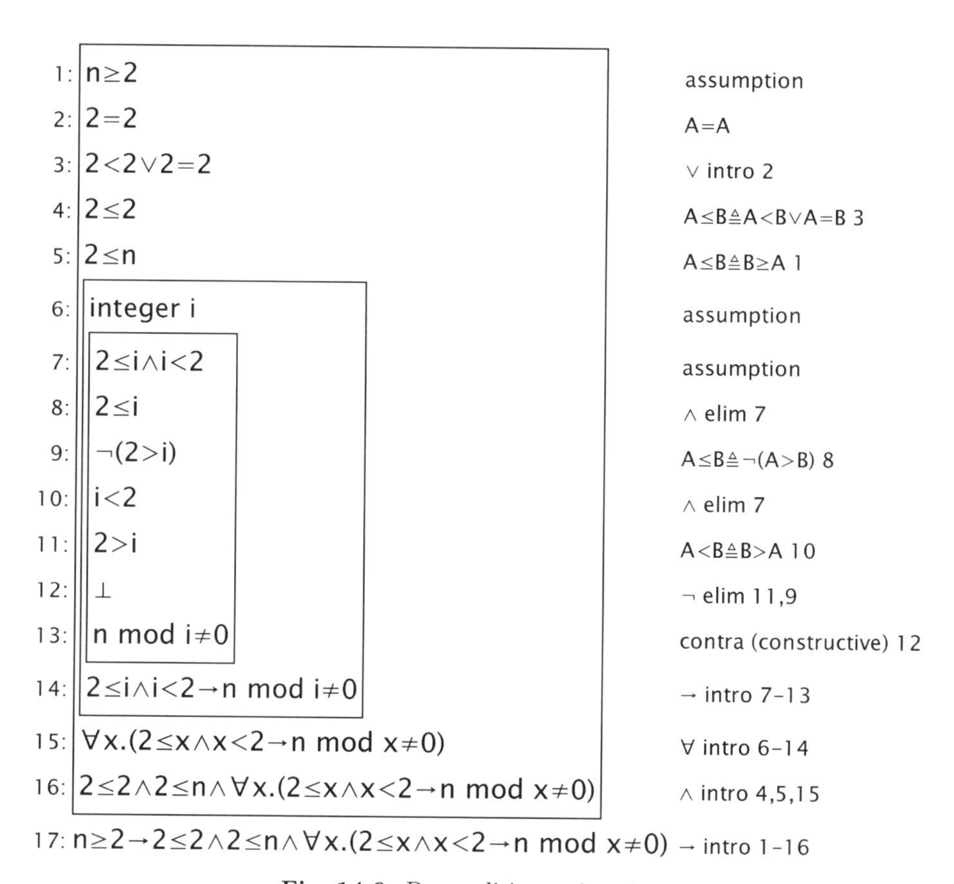

Fig. 14.6 Precondition to invariant

1:	$2\leq i \wedge i \leq n \wedge \forall x.(2\leq x \wedge x<i \rightarrow n \bmod x \neq 0)$	assumption
2:	$2\leq i$	$\wedge$ elim 1
3:	$i \neq 0$	obviously, from 2
4:	$2\leq i \wedge i \leq n \wedge \forall x.(2\leq x \wedge x<i \rightarrow n \bmod x \neq 0) \rightarrow i\neq 0$	$\rightarrow$ intro 1-3

Fig. 14.7 Remaindering won't crash

over integers (in fact the assumption on line 6 is never used, but we still need the privacy condition, of course). Once you spot the contradiction inherent in line 7 it's only a matter of pummelling the inequalities into shape. It doesn't even need an 'obviously' step.

14.5.3 The loop guard won't crash. The definedness antecedent of the while rule is $I \rightarrow (E \text{ computes})$, and the loop guard formula E in the primacy example

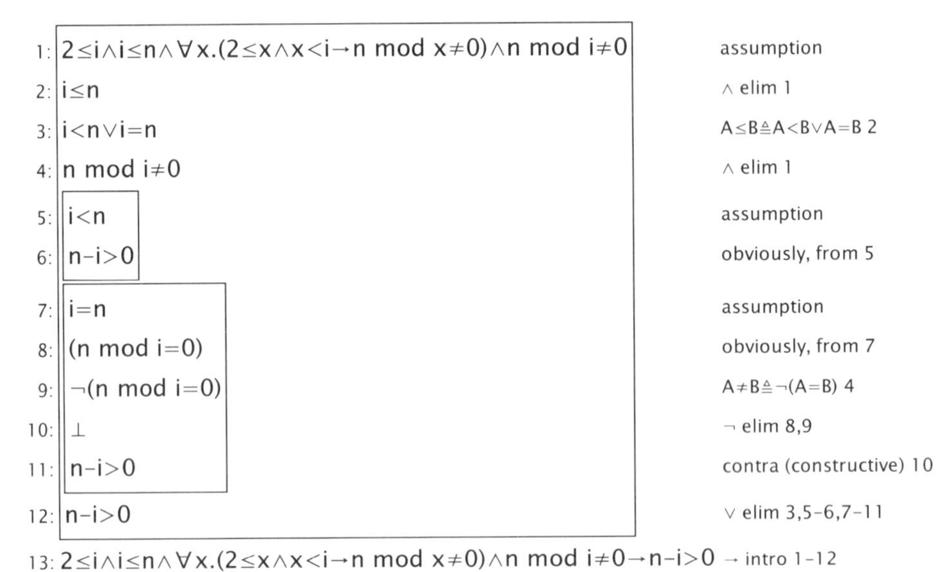

Fig. 14.8 Measure isn't exhausted

is n mod $i = 0$. In the interpretation of finite-width-numeral arithmetic encoded in this version of Hoare logic, that formula will evaluate correctly provided $i \neq 0$. Jape hides all the deductions that lead to the proof obligation on line 2 of Fig. 14.5. The proof of the obligation is trivial, and shown in Fig. 14.7. (Hoare-logic Jape is set up to do ∧ elim and ∧ intro without fuss, so it hides the steps which extract line 2 of Fig. 14.7 from line 1.)

14.5.4 The measure is greater than zero on every execution. Line 4 of Fig. 14.5 is generated directly from line 7 by the while rule. It requires that when we have the invariant and the guard — i.e. at the start of every execution of the loop body — the measure is greater than zero. In more careful language "the invariant and the guard together imply that the measure isn't exhausted".

The problem, once I've unified `_M` with $n - i$, is stated on line 13 of Fig. 14.8. The matter is very arithmetical. On the left-hand side we have $i \leq n$, which is equivalent to $i < n \vee i = n$. But we also have n mod $i \neq 0$, which tells us — this is the insight we need — that $n \neq i$ (by contradiction: if $n = i$, the remainder on division would certainly be zero). So we can be sure that $i < n$, which after some simple manipulation is surely the same thing as $0 < n - i$, which is the right-hand side written the other way round.

Jape manages most of that argument. It shows how the contradiction works, where the clever bit of arithmetic reasoning is appealed to, and where the arithmetic manipulations that it can't do ought to go.

1:	2≤i∧i≤n∧∀x.(2≤x∧x<i→n mod x≠0) ∧n mod i≠0∧n−i=Km	assumption
2:	n−i=Km	∧ elim 1
3:	n−(i+1)<n−i	obviously
4:	n−(i+1)<Km	equality-substitution 2,3
5:	2≤i∧i≤n∧∀x.(2≤x∧x<i→n mod x≠0) ∧n mod i≠0∧n−i=Km→n−(i+1)<Km	→ intro 1-4

Fig. 14.9 Measure reduces at each step

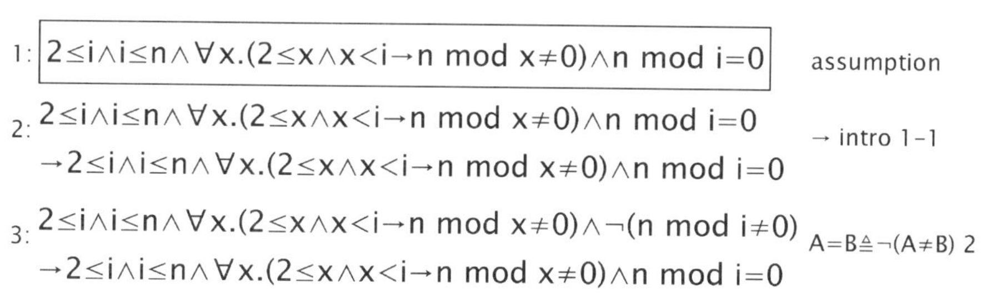

Fig. 14.10 Termination gives postcondition

14.5.5 The measure reduces. Line 6 in Fig. 14.5 requires that $i := i + 1$, the body of the loop, reduces M: if it starts equal to some arbitrary but fixed integer Km, it must finish up less than Km (Km is private to the box containing line 6, so we can't cheat by defining it outside). In careful language "the invariant and the guard together imply that the loop body reduces the measure" — though in this example the invariant and the guard don't need to come into it. It's pretty easy to prove: if $n - i = Km$ then the right-hand side of the implication can be rewritten, replacing Km with its equivalent, as $n - (i + 1) < n - i$. After a bit of algebraic manipulation that would reduce to $0 < 1$, but it's pretty close to obvious as it stands.

The Jape proof, shown in Fig. 14.9, is an exercise in button-pushing.

14.5.6 We have the right postcondition. In stating the postcondition of the program (see line 10 of Fig. 14.5) I wrote $n \bmod i = 0$; the while rule, on the other hand, requires the negation of the loop condition, $\neg(n \bmod i \neq 0)$. Classically the two formulae are equivalent, but they aren't identical: therefore Jape introduces the consequence step on line 9 and the verification condition on line 8. Formally we have to prove that "the invariant and the negation of the guard imply the postcondition": in this case it's a trivial matter, a single-step

Line	Formula	Justification
1:	$2 \leq i \wedge i \leq n \wedge \forall x.(2 \leq x \wedge x < i \rightarrow n \bmod x \neq 0) \wedge n \bmod i \neq 0$	assumption
2:	$2 \leq i$	$\wedge$ elim 1
3:	$i \leq n$	$\wedge$ elim 1
4:	$i < n \vee i = n$	$A \leq B \triangleq A < B \vee A = B$ 3
5:	$\forall x.(2 \leq x \wedge x < i \rightarrow n \bmod x \neq 0)$	$\wedge$ elim 1
6:	$n \bmod i \neq 0$	$\wedge$ elim 1
7:	$2 \leq i+1$	obviously, from 2
8:	$i < n$	assumption
9:	$i+1 \leq n$	$A+1 \leq B \triangleq A < B$ 8
10:	$i = n$	assumption
11:	$(n \bmod i = 0)$	obviously, from 10
12:	$\neg(n \bmod i = 0)$	$A \neq B \triangleq \neg(A = B)$ 6
13:	$\bot$	$\neg$ elim 11,12
14:	$i+1 \leq n$	contra (constructive) 13
15:	$i+1 \leq n$	$\vee$ elim 4,8-9,10-14
16:	$\forall x.(2 \leq x \wedge x < i+1 \rightarrow n \bmod x \neq 0)$	$\forall x.(A \leq x \wedge x < B \rightarrow P(x)), P(B) \ldots$ 5,6
17:	$2 \leq i+1 \wedge i+1 \leq n \wedge \forall x.(2 \leq x \wedge x < i+1 \rightarrow n \bmod x \neq 0)$	$\wedge$ intro 7,15,16
18:	$2 \leq i \wedge i \leq n \wedge \forall x.(2 \leq x \wedge x < i \rightarrow n \bmod x \neq 0) \wedge n \bmod i \neq 0 \rightarrow 2 \leq i+1 \wedge i+1 \leq n \wedge \forall x.(2 \leq x \wedge x < i+1 \rightarrow n \bmod x \neq 0)$	$\rightarrow$ intro 1-17

Fig. 14.11 Invariant is preserved

equality-substitution of a single formula followed by a trivial $\rightarrow$ intro, shown in Fig. 14.10.

14.5.7 The invariant is preserved. This is the big one. Line 4 of Fig. 14.5 expresses the condition that "the invariant and the guard together imply that the loop body preserves the invariant". The verification condition it generates is line 18 of Fig. 14.11.

Most of the proof is straightforward arithmetic. Line 11 needs insight, but we've already spotted it in the measure proof (Fig. 14.8). Line 16 is deduced from lines 5 and 6 via an instance of a **loop-rolling** theorem, whose proof is shown in Fig. 14.12. (The distinctness provisos on the theorem require that A and B are loop-bounds formulae independent of x, which is clearly the case in this example.)

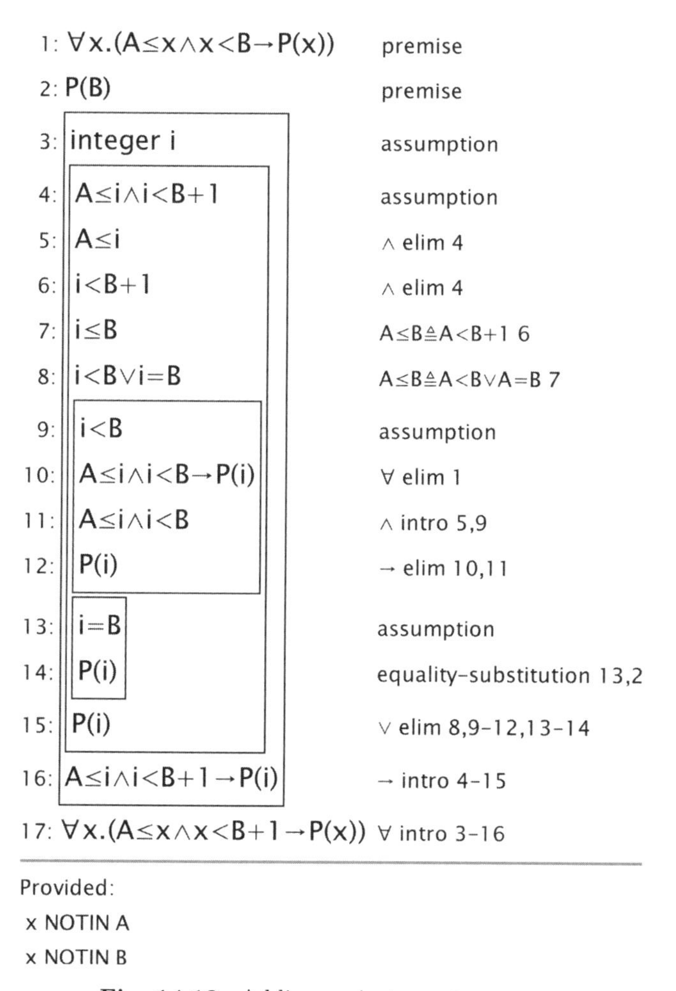

Fig. 14.12 Adding an instance to a range

14.5.8 And finally! We know that the program does terminate (Figs. 14.8 and 14.9), provided we have $n \geq 2$ to start with (Fig. 14.6); we know that when it terminates $2 \leq i \leq n$ (Figs. 14.11 and 14.10), i.e. i really is a number in the range $2..n$; we know also that $\forall x(2 \leq x < i \rightarrow n \bmod i \neq 0)$, i.e. no number in the range $2..i-1$ is a factor of n; we know that $n \bmod i = 0$, i.e. i **is** a factor of n.

If the program terminates with $i = n$ then no number in the range $2..n-1$ is a factor of n, and that's a pretty fair approximation to the notion of 'prime number'. If on the other hand $i \neq n$ then we've found a number in the range $2..n-1$ which is a factor of n. The instruction $prime := (i = n)$ will surely assign the correct Boolean value to *prime*; it seems to me, waving my hands, that I

Line	Formula	Justification
1:	$2 \le i \wedge i \le n \wedge \forall x.(2 \le x \wedge x < i \rightarrow n \bmod x \ne 0) \wedge n \bmod i = 0$	assumption
2:	$2 \le i$	$\wedge$ elim 1
3:	$i \le n$	$\wedge$ elim 1
4:	$i < n \vee i = n$	$A \le B \triangleq A < B \vee A = B$ 3
5:	$\forall x.(2 \le x \wedge x < i \rightarrow n \bmod x \ne 0)$	$\wedge$ elim 1
6:	$n \bmod i = 0$	$\wedge$ elim 1
7:	$i = n$	assumption
8:	$\forall y.(2 \le y \wedge y < n \rightarrow n \bmod y \ne 0)$	equality-substitution 7,5
9:	$i = n \rightarrow \forall y.(2 \le y \wedge y < n \rightarrow n \bmod y \ne 0)$	$\rightarrow$ intro 7-8
10:	$\forall y.(2 \le y \wedge y < n \rightarrow n \bmod y \ne 0)$	assumption
11:	$i < n$	assumption
12:	$2 \le i \wedge i < n \rightarrow n \bmod i \ne 0$	$\forall$ elim 10
13:	$2 \le i \wedge i < n$	$\wedge$ intro 2,11
14:	$n \bmod i \ne 0$	$\rightarrow$ elim 12,13
15:	$\neg(n \bmod i = 0)$	$A \ne B \triangleq \neg(A = B)$ 14
16:	$\bot$	$\neg$ elim 6,15
17:	$i = n$	contra (constructive) 16
18:	$i = n$	assumption
19:	$i = n$	$\vee$ elim 4,11-17,18-18
20:	$\forall y.(2 \le y \wedge y < n \rightarrow n \bmod y \ne 0) \rightarrow i = n$	$\rightarrow$ intro 10-19
21:	$(i = n \rightarrow \forall y.(2 \le y \wedge y < n \rightarrow n \bmod y \ne 0)) \wedge (\forall y.(2 \le y \wedge y < n \rightarrow n \bmod y \ne 0) \rightarrow i = n)$	$\wedge$ intro 9,20
22:	$i = n \leftrightarrow \forall y.(2 \le y \wedge y < n \rightarrow n \bmod y \ne 0)$	$\leftrightarrow$ intro 21
23:	$2 \le i \wedge i \le n \wedge \forall x.(2 \le x \wedge x < i \rightarrow n \bmod x \ne 0) \wedge n \bmod i = 0 \rightarrow i = n \leftrightarrow \forall y.(2 \le y \wedge y < n \rightarrow n \bmod y \ne 0)$	$\rightarrow$ intro 1-22
24:	$\{i = n \leftrightarrow \forall y.(2 \le y \wedge y < n \rightarrow n \bmod y \ne 0)\}(prime := i = n)\{prime \leftrightarrow \forall y.(2 \le y \wedge y < n \rightarrow n \bmod y \ne 0)\}$	variable-assignment
25:	$\{2 \le i \wedge i \le n \wedge \forall x.(2 \le x \wedge x < i \rightarrow n \bmod x \ne 0) \wedge n \bmod i = 0\}(prime := i = n)\{prime \leftrightarrow \forall y.(2 \le y \wedge y < n \rightarrow n \bmod y \ne 0)\}$	consequence(L) 23,24

Provided:
DISTINCT i, n, prime

Fig. 14.13 Prime numbers really are detected

have proved that the naive algorithm really does assess the primacy of integers 2 and above.

But we can do better than hand-waving. The formal definition of primacy is $\neg\exists x(2 \leq x < n \land n \bmod x = 0)$, which is classically equivalent to $\forall x(2 \leq x < n \rightarrow n \bmod x \neq 0)$. A formal proof that $prime := (i = n)$ is adequate is shown in Fig. 14.13. Not an 'obviously' step in it, and lines 11 to 16 explain just why you can't have both $i < n$ and also n prime.

14.5.9 Is it worth it? Five arithmetical proofs to show that the loop in the prime-finding program terminates provided that it doesn't crash; one to show that it doesn't crash; one more to show that we always get just the result we asked for. It's not a formal argument, because there are six 'obviously' steps. So: does semi-formalism pay off?

I think it does. The proofs are imperfect, but imperfection has been chased into their farthest corners. They reinforce and illuminate our informal understanding. Formality helps informality survive and keeps it honest. It **is** worth it.

14.5.10 These proofs are classical. Although the proofs don't use classical contradiction steps, they are resolutely classical in their treatment of arithmetic. In several places, for example, they derive $A = B$ from $\neg A \neq B$. A constructivist would object that we might be sure we can't prove $A \neq B$, without necessarily being able to prove $A = B$.

In the places where non-constructive steps used, though, we are making **decidable** comparisons between the values of variables and constants, or **computable** combinations of such values, so the objection doesn't have much force. A constructivist could provide a proof of the same conclusions following the same course, just with rather a lot of extra fuss. I would say that these proofs don't raise any constructivist hackles. Nevertheless, in terms of the steps that they use, they are classical.

14.6 Yet more arithmetic

We're all brainwashed in first school to slavishly follow the methods of Abu Ja'far Muhammad ibn Musa al-Khwarizmi, the celebrated eponymous inventor of the **algorithms** they teach us to use for addition, subtraction, multiplication and division. His inventions are indeed marvellous, and he deserves all of his enduring fame (he's really worth looking up in the library or on the internet), but they aren't the last word.

It's easy to forget, after such an education, that multiplication is no more than repeated addition and that integer division is counted repeated subtrac-

```
...
1: {i=Ki∧j=Kj∧i≥0}(k:=0){i≥0∧k+i×j=Ki×Kj}
...
2: {i≥0∧k+i×j=Ki×Kj∧i≠0}(k:=k+j;i:=i-1){i≥0∧k+i×j=Ki×Kj}
...
3: i≥0∧k+i×j=Ki×Kj∧i≠0→_M>0
4: | integer Km                                                    | assumption
   | ...                                                           |
5: | {i≥0∧k+i×j=Ki×Kj∧i≠0∧_M=Km}(k:=k+j;i:=i-1){_M<Km}            |
6: {i≥0∧k+i×j=Ki×Kj}while i≠0 do k:=k+j;i:=i-1 od                 while 2,3,4-5
   {i≥0∧k+i×j=Ki×Kj∧¬(i≠0)}
...
7: i≥0∧k+i×j=Ki×Kj∧¬(i≠0)→k=Ki×Kj
8: {i≥0∧k+i×j=Ki×Kj}while i≠0 do k:=k+j;i:=i-1 od{k=Ki×Kj}  consequence(R) 6,7
9: {i=Ki∧j=Kj∧i≥0}(k:=0){i≥0∧k+i×j=Ki×Kj}                         Ntuple 1,8
   while i≠0 do k:=k+j;i:=i-1 od{k=Ki×Kj}
```

Provided:
DISTINCT i, j, k

Fig. 14.14 The beginning of a proof of repeated-addition multiplication

tion. Once you remember, you can write computer programs that exploit those meanings, and prove that they work.

For example, multiplication. Start with 0; add j exactly i times; you must finish up with $i \times j$.

$$\begin{array}{l} \{i = Ki \wedge j = Kj \wedge i \geq 0\} \\ \quad k := 0; \\ \{i \geq 0 \wedge k + i \times j = Ki \times Kj\} \\ \quad \textsf{while } i \neq 0 \textsf{ do } k := k + j;\ i := i - 1 \textsf{ od} \\ \{k = Ki \times Kj\} \end{array} \tag{14.2}$$

The proof starts as shown in Fig. 14.14. The loop measure is i. The proof obligations are really easy to prove formally, modulo a bit of cancellation in number-algebra formulae that Jape isn't really up to.

For another example, division. Start with i; keep subtracting j until the next subtraction would go negative; the quotient is the number of subtractions you managed to do, and the remainder is the number you decided not to reduce further.

$$\{i = Ki \wedge j = Kj \wedge i \geq 0 \wedge j > 0\}$$
$$kq := 0; kr := i;$$
$$\{j = Kj \wedge j > 0 \wedge kr \geq 0 \wedge kq \times j + kr = Ki\} \quad (14.3)$$
$$\textsf{while } kr \geq j \textsf{ do } kr := kr - j;\ kq := kq + 1 \textsf{ od}$$
$$\{kq \times Kj + kr = Ki \wedge 0 \leq kr < Kj\}$$

This time the measure is kr, the number being reduced towards the remainder. It reduces on each step provided that $j > 0$ (that shows why you can't divide by zero: you'd have to loop for ever).

14.7 Can we improve on al-Khwarizmi?

We can't all be geniuses at his level, but we can adapt al-Khwarizmi's digit-by-digit method to modern programming. Every lazy schoolchild knows that next to multiplying by zero, multiplying by ten and dividing by ten are the easiest operations to apply to a decimal numeral. Add a trailing zero is $(\times 10)$, cross out the last digit is $(\div 10)$. You should now realize that $(\bmod 10)$ is just as easy: simply pick out the last digit.

Then it's possible to see that the schoolroom multiplication algorithm is based on the identity $(A \div 10) \times 10 + A \bmod 10 = A$:

$$\begin{aligned} A \times B &= ((A \div 10) \times 10 + A \bmod 10) \times B \\ &= ((A \div 10) \times B) \times 10 + (A \bmod 10) \times B \end{aligned} \quad (14.4)$$

— multiply B by the right-hand digit; add that to the result of multiplying the previous digits by B, shifted one place left. Here's an example.[1]

$$\begin{array}{ccccccc} & & & 4 & 9 & 7 & 2 \\ & & \times & & 6 & 8 & 3 \\ \hline 2 & 9 & 8 & 3 & 2 & & \\ & 3 & 9 & 7 & 7 & 6 & \\ & & 1 & 4 & 9 & 1 & 6 \\ \hline 3 & 3 & 9 & 5 & 8 & 7 & 6 \end{array} \quad (14.5)$$

14916 is 4972×3; the two lines above it are 4972×68, shifted one place left (i.e., multiplied by 10). 39776 is 4972×8; the line above it is 4972×6, shifted a further place left. Add up all the single-digit multiplications and you get the right answer.

I don't know whether al-Khwarizmi recognized the possibility of numeral-bases other than 10. Whether he did or not, we can exploit them. Computer hardware works with binary numerals nowadays, and that means that $(\div 2)$, $(\bmod 2)$

[1] You may have been taught to put dots or zeros where I've put spaces at the end of the third and fourth lines. The difference is immaterial.

and $(\times 2)$ are very simple and fast operations on non-negative numerals: shift right is $(\div 2)$, mask with 1 is (mod2), and shift left is $(\times 2)$. The old digit-by-digit trickery works better than ever, because single-digit multiplication is so easy: $(A \bmod 2) \times B$ is if $A \bmod 2 = 1$ then B else 0 fi. After a bit of thought, it's easy to program the whole mechanism:

$$
\begin{array}{l}
\{i = Ki \wedge j = Kj \wedge i \geq 0\} \\
\quad k := 0; \\
\{i \geq 0 \wedge k + i \times j = Ki \times Kj\} \\
\quad \textsf{while } i \neq 0 \textsf{ do} \\
\qquad \textsf{if } i \bmod 2 = 1 \textsf{ then } k := k + j \textsf{ else skip fi}; \\
\qquad i := i \div 2; j := j \times 2 \\
\quad \textsf{od} \\
\{k = Ki \times Kj\}
\end{array}
\tag{14.6}
$$

The proof works with i as a measure, but that's a massive over-estimate. This is **fast** integer multiplication: the number of times round the loop is the length of the binary numeral which represents i (by contrast, (14.2) takes time proportional to the value of i, which is usually much, much larger). But never mind: the while rule is happy to use an overestimating measure.

The proof is still tricky, though, because you have to be careful to notice that $i \div 2 < i$ only when $i > 0$. (It's very easy to miss that point, which illustrates why programming is so darned difficult and why the crackers will continue to find ways round our defences at least until we find ways of plugging them formally.)

Fast integer division — I was taught it as 'long division' — is more intricate. If $q \times B + r = A \div 10$, then $q \times B \times 10 + r \times 10 = (A \div 10) \times 10 = A - A \bmod 10$, and $(q \times 10) \times B + r \times 10 + A \bmod 10 = A$. So:

- if $A < B$, quotient is 0, remainder is A;
- otherwise, divide $A \div 10$ by B; multiply quotient and remainder by 10; add $A \bmod 10$ to the remainder; subtract B as many times as you can from the new remainder, adding 1 to the quotient each time.

Al-Khwarizmi understood that, though he might have put it differently.[2] I've relied on a similar mechanism, using 2 instead of 10 and manipulating B rather than A:

- if $A < B$, quotient is 0, remainder is A;
- otherwise, divide A by $B \times 2$; multiply quotient by 2; subtract B as many times as you can from the remainder, adding 1 to the quotient each time.

[2] The only surviving version of his work is a book in mediaeval Latin. I have to confess that I haven't read it.

The program starts with a repeated shift left (i.e. repeated $\times 2$) which is accounted for in variable kc, followed by a repeated shift right (repeated $\div 2$) of exactly the same amount.

$$
\begin{array}{l}
\{i = Ki \wedge j = Kj \wedge i \geq 0 \wedge j > 0\} \\
\quad kq := 0; kr := i; kc := 0; \\
\{j = Kj \times 2^{kc} \wedge j > 0 \wedge kq \times j + kr = Ki \wedge 0 \leq kr\} \\
\quad \textsf{while } j \leq kr \textsf{ do } j := j \times 2; kc := kc + 1 \textsf{ od}; \\
\{j = Kj \times 2^{kc} \wedge j > 0 \wedge kq \times j + kr = Ki \wedge 0 \leq kr < j\} \\
\quad \textsf{while } kc \neq 0 \textsf{ do} \\
\qquad j := j \div 2; kc := kc - 1; kq := kq \times 2; \\
\qquad \textsf{if } j \geq kr \textsf{ then } kr := kr - j; kq := kq + 1 \textsf{ else skip fi} \\
\quad \textsf{od} \\
\{kq \times Kj + kr = Ki \wedge 0 \leq kr < Kj\}
\end{array}
\tag{14.7}
$$

This problem is also available in Jape. As usual the proof depends on the precondition $j > 0$: I'll leave you to spot where it's needed. I'll also leave you to struggle with the problem of showing why, when $kr < j$ and $j = Kj \times 2^{kc} \wedge kc > 0$, it follows that $kr - j \div 2 < j \div 2$.

14.8 None of these programs has been tested

It's traditional, in Hoare-logic presentations, to boast that your examples have been proved but have never been tested on a machine. That's the safest way to program, but programming remains programming and you can still fall flat on your face. If the pre- and post-condition formulae don't exactly capture what you want the program to do, it will do what you proved instead of what you wanted. If the bits of unformalized arithmetic you waved your hands over are not quite as obvious as you hoped, nasties can happen. But if you are careful, you'll make fewer of those kinds of mistakes than you would make mistakes in 'normal' programming.

In the end, though, all you can prove is a meta-implication: if the specification describes what you want, then the program will do what you want. Mis-specification doesn't get you the program you hoped for. The gap between the world we live in and the logical worlds we invent, formal or informal, is as wide as ever.

15 Arrays

The proof of the primacy checker in Chapter 14 is fun in a daft sort of way, but it's a bit too much about arithmetic. The arithmetic algorithms are more fun and even a bit useful, but they are completely about arithmetic. Although arithmetic lies in wait in the depths of every Hoare-logic proof, we needn't provoke it so. Programs which use arrays — for searching, sorting, and so on — are challenges with are both more realistic and less obstinately arithmetical.

15.1 Two extra formulae and an extra instruction

Programs in the language of Chapters 13 and 14 include formulae made up of integer and Boolean constants, integer and Boolean variable names, and integer and Boolean operators.

In this chapter I add **array variables** to that mix: I use a, b and c in my examples. I allow **array-element formulae** of the form $a[E]$, where a is a an array variable and E is an integer-valued formula, naming the Eth element of the sequence contained in variable a.

There's an extra instruction as well: $a[E] := F$ replaces the Eth element of the a-sequence with the value of formula F. You can write simple assignments like $a[2] := 17$ and horrendously complicated ones like $a[a[i]] := i + a[a[j-1]+1]$.

Because definedness matters so much when accessing arrays, there's also a formula describing the size of an array: $\text{length}(a)$ describes the length of the sequence stored in variable a. You can't extend or reduce that length by assignment.

As before, I shan't be bothering with typing or declarations. I shall use names i, j and k for integer variables, a, b and c for array variables, and the kinds of sequences stored in arrays will be evident from context.

Pronunciation $a[E]$ is pronounced "a sub E" to reflect its mathematical heritage. Mathematicians use subscripting (writing below) as, for example, in V_i which describes the ith element of a vector V or $M_{i,j}$ which picks out an element of a matrix M. Computer scientists, using paper tape and punched cards long before they had keyboards, colour screens and multifont GUIs, had to use square brackets instead of writing below the line, but they hung on to the mathematical nomenclature.

For that reason E in $a[E]$ is often called a subscript expression, and the operation of selecting an element subscripting. Because I'm interested in the calculation which underpin programming (and because I'm proud that I once worked for the University of Manchester) I shall call the formula the **index** and the operation **indexing**.

15.2 Address arithmetic and the array-bounds problem

Arrays, inside a modern computer, are contiguous sequences of memory locations — i.e. contiguous sequences of variables. Array-element formulae, like $a[3]$, pick out one of the variables by using simple integer arithmetic: take the address of the first element in the sequence and add the index. Thus, for example, $a[3]$ names the variable at address $a + 3$ — reading a just as a simple address.[1] In C it works just like that, and in other languages — even Java — it's hardly different.

It's a neat trick, but it gives us a problem. Given an array (i.e. an address) a you can pick an index E which makes $a[E]$ refer to any location in memory that you want it to, including even a location outside the array a itself. That might sound relatively harmless, but it follows that the assignment $a[E] :=$ N can, by using a carefully selected index E, write any N that you like into whatever location you choose. That is, you can do arbitrary changes to the memory, sometimes including the memory that holds your program, just by using array-element indexing — and the machine won't stop you.

This is the **array-bounds** problem, and at present it's the most frequently used loop-hole used by the crackers who wriggle into other people's computers without permission. Some programming languages — for example, C — allow their users to do their own address arithmetic, so you can play addressing tricks on yourself and drill your own bigger loop-holes to let the crackers in more easily. More constrained languages, like the one used in this book, hide the address arithmetic that make arrays work, and their programs check every indexing $a[E]$ to make sure that the index formula keeps within the bounds of the array, crashing if ever it strays.

15.3 A formal treatment of array-element assignment

15.3.1 Definedness. If programs that use arrays are to terminate, definedness of $a[E]$ is an important issue. E must compute in finite-width-numeral

[1] Actually it's slightly more complicated than that. Modern computers divide their store into 'bytes' or 'octets' of 8 bits each, and integer variables are short sequences of bytes: 2 or 4 or 8 bytes, usually. So an array of integer variables is actually an array of 2-byte or 4-byte or 8-byte sequences, and $a[3]$ is at address $a+6$ or $a+12$ or $a+24$ accordingly. But the principle is the same: **simple, fast** integer arithmetic gives us array indexing.

arithmetic, and its value must be within the indexing bounds of the sequence identified by a.

$$\textbf{Definition 15.1}\ \frac{\begin{array}{cc}\vdots & \vdots \\ E\ \mathit{computes} & 0 \leq E < \text{length}(a)\end{array}}{a[E]\ \mathit{computes}}$$

15.3.2 Aliasing. The proofs in chapters 13 and 14 needed provisos to ensure that you couldn't prove nonsense. You can prove the theorem $\{j=2\}\, i:=1\, \{j=2\}$, for example, for arbitrary distinct variables i and j — but you can't prove $\{k = 2\}\, k := 1\, \{k = 2\}$, even though it looks like an instance of the theorem produced by replacing both i and j by k. The statement of the theorem has to prohibit that sort of nonsense, and hence the proviso DISTINCT i, j in the Jape theorem and proof.

This is a sight of the famous problem of **aliasing** in Hoare logic. It's quite possible in many programming languages to arrange that two variable names refer to the same memory location (and when it isn't possible, you can usually do something just as horrible by messing about with pointers or references). It's confusing when it happens, and it stops the variable-assignment axiom working unless you hit it on the head with a device like Jape's DISTINCT proviso.

Substitution doesn't work unless you can tell what is an alias and what is not. Array-element aliasing is a consequence of address arithmetic. The element formulae $a[E]$ and $a[E']$ are aliases — i.e. they refer to the same array element, the same memory location — just when $E = E'$: which is to say, rather easily and rather often. Aliasing, which I hit on the head with a distinctness proviso in Chapter 13, has popped up again like a fairground whack-a-mole. But array-element aliasing, unlike variable aliasing, isn't a rare occurrence that we can easily stun. If we tried to mimic the variable-assignment axiom directly, reading $a[E] := F$ as an instruction to substitute F for $a[E]$ to make a precondition, aliasing would too often make nonsense of the result. We need another approach.

15.3.3 Arrays as single-variable sequences. The solution — the only possible solution in Hoare logic — is to treat arrays as if they were single variables. Instead of seeing an array as a collection of separate variables, we imagine that an array variable contains a single sequence value, just as an integer variable contains a single integer value. If a contains the five-element sequence $\langle 3, 5, 17, -6, 32\rangle$, for example, $a[3] = -6$. Then the effect of $a[3] := 42$ is to assign to variable a the modified sequence $\langle 3, 5, 17, 42, 32\rangle$.

I number sequences starting with 0 — like Java, C and C++, but unlike Pascal, Algol and Fortran. An n-element sequence can be indexed by any integer from 0 to $n - 1$.

This idea is a reasonable extension of the treatment of integer variables. The program $i := i + 1$, for example, must pick up a number from integer variable i, add 1 to that number, and write back the result. In just the same way the program $a[3] := 42$ must pick up a sequence from array variable a, modify it so that its fourth element is 42, and write back the result.

It isn't necessarily an inefficient idea either. A computer can cut corners by only picking up the bit of the sequence it needs to modify or by overwriting part of the sequence in memory without picking up anything at all. Since the effect is the same in either case as if there was a variable containing the sequence, the implementation of arrays can continue to be exactly what it always was. But our formal treatment of arrays has to deal with them as if they were single variables — a plausible and useful fiction in the examples I shall consider.

15.3.4 Updatable-sequence notation. What's the effect of $a[3] := 42$ in general: not on some particular sequence but supposing that a contains an arbitrary fixed sequence Ka? The answer is: provided that Ka has at least four elements, put a sequence into a which is exactly like Ka except that at position 3 it has 42. We write that sequence as $Ka \oplus 3 \mapsto 42$.

These updatable-sequence formulae can be indexed. You would expect that $(Ka \oplus 3 \mapsto 42)[3]$ must be 42, and it is. $(Ka \oplus 3 \mapsto 42)[2]$, on the other hand, doesn't care what's at index 3 and therefore reduces to $Ka[2]$.

To index an updatable-sequence formula $A \oplus E \mapsto F$ you always have to prove an equality or an inequality.

Definition 15.2
$$\frac{\begin{array}{c}\vdots\\ E = E'\end{array}}{(A \oplus E \mapsto F)[E'] = F}\ \mathit{array\text{-}indexing}(R)$$

Definition 15.3
$$\frac{\begin{array}{c}\vdots\\ E \neq E'\end{array}}{(A \oplus E \mapsto F)[E'] = A[E']}\ \mathit{array\text{-}indexing}(L)$$

15.3.5 The array-element assignment axiom.

Definition 15.4

$$\{\ (a[E]\ \mathit{computes}) \wedge (F\ \mathit{computes})\ \wedge B^{a}_{a \oplus E \mapsto F}\ \}\ a[E] := F\,\{B\}$$

Don't be scared of the formulae in this axiom: $B^{a}_{a \oplus E \mapsto F}$ simply means "a copy of B in which every occurrence of a has been replaced by $a \oplus E \mapsto F$". Although the replacement formula is more ferocious, and the definedness conditions more demanding, it just describes a single-variable assignment to a, so it's nothing more than a specialized version of the variable-assignment axiom.

15.3.6 Definedness: cooling down. The condition "$a[E]$ computes" means no more than (E computes) $\wedge\, 0 \leq E < \text{length}(a)$. A precondition must imply this definedness condition if $a[E] := F$ is not to crash while calculating the index value or when using it to point to an element of array a. That means you can't expect to prove $\{\top\}\, a[i] := 0\, \{a[i] = 0\}$: $\top$ holds in any state; so it holds when $i = -1$, in particular; but in that state the program crashes. We have a counter-example: there can't be a proof.

What about $\{a[i] = 3\}\, a[i] := 0\, \{a[i] = 0\}$? Necessarily, if $a[i] = 3$, i must be within the indexing range of a, because $a[i]$ isn't defined outside that range. We know that $a[i] = 3 \rightarrow (a[i]$ computes); the assignment won't crash.

We're considering implications here, and implications always muddy the waters. What about a precondition $a[i] = i \div 0$? Program variables and array elements can only hold finite values, so if it's true then I'm a banana — but that just lets the cunning uncle in as usual, and we're forced to agree that $a[i] = i \div 0 \rightarrow (a[i]$ computes).

It's all rather subtle. $a[E] = 3$ implies $0 \leq E < \text{length}(a)$ — E is within index bounds — but it doesn't imply E computes, because that claim is about finite-width-numeral computability, and E might be more complicated than that. It could contain some infinity-naughtiness: $a[\textsf{if } i \div 0 \neq 3 \textsf{ then } 0 \textsf{ else } -1 \textsf{ fi}]$, for example, is always $a[0]$, therefore always in bounds unless a holds the empty sequence — but you can't calculate the index value using finite arithmetic. The element may be ok even when the index formula is not.

Jape tries its best to keep up with all this. To reduce noise in your proof it minimizes the number of occurrences of $0 \leq E$ and $E < \text{length}(a)$ in a precondition, it exploits the fact that $\text{length}(A \oplus E \mapsto F) = \text{length}(A)$, and it lets you deduce the definedness of array elements which appear in hypothesis equalities and inequalities.

15.3.7 Definedness: not so fast! It's nice to be able to deduce that i is in indexing bounds from $a[i] = 2$, but there's a downside to this meta-arithmetic. If $A = A$ is an axiom then we always have $a[E] = a[E]$, from which we can deduce $0 \leq E < \mathit{length}(a)$ for no matter what a and what E. That's absurd: we can pick -1 as our E, for example, and deduce $0 \leq -1$ and from there we can find a contradiction, because it's surely axiomatic that $-1 < 0$.

So: if $A = A$ is an axiom, we can deduce a contradiction and our logic is unsound. On the other hand, we do want to conclude $A = A$ quite often, if for nothing else than to simplify updatable-sequence formulae using Definition 15.2. No problem! If it can't be an axiom we can still have a rule:

...

1: a[i]=2→(a⊕i↦a[i]+1)[i]=3∧0≤i∧i<length(a)

2: {(a⊕i↦a[i]+1)[i]=3∧0≤i∧i<length(a)}(a[i]:=a[i]+1){a[i]=3} array-element-assignment

3: {a[i]=2}(a[i]:=a[i]+1){a[i]=3} consequence(L) 1,2

Provided:
DISTINCT a, i

Fig. 15.1 Incrementing an array element: verification condition extracted

$$\vdots$$

Definition 15.5 $\dfrac{A\ defined}{A = A}\ arithmetic\ identity$

'A defined' means no more than 'A doesn't contain any out-of-bounds array accesses'. That's the sort of thing that a proof tool like Jape can deal with. (It's also the sort of verification condition that it's easy to miss in an informal proof!)

If we do have $a[-1] = a[-1]$ — if it's a premise, for example — then we're in cunning-uncle territory. We can deduce a contradiction, and therefore any conclusion we like. But, of course, such a cunning proof will be useless for all the usual reasons.

A similar difficulty arises with $A < B \vee A = B \vee A > B$, which needs both ($A$ defined) and (B defined) as antecedents.

15.4 Simple array element assignment examples

15.4.1 Incrementing an element of a sequence. Incrementing an integer array element can't be harder than incrementing an integer variable, surely? Well ... a little bit harder, because you have to be sure you don't go outside the array bounds.

Consider, for example, $\{a[i] = 2\}\ a[i] := a[i] + 1\ \{a[i] = 3\}$. This one's valid: the business end of the precondition calculated by the array-element assignment axiom is $(a[i] = 3)^{a}_{(a \oplus i \mapsto a[i]+1)}$, which is $(a \oplus i \mapsto a[i] + 1)[i] = 3$, which simplifies immediately (Definition 15.2) to $a[i] + 1 = 3$, which is surely the same as $a[i] = 2$. And we can forget about the definedness preconditions, because certainly i computes and from $a[i] = 2$ we know that $a[i]$ and therefore $a[i] + 1$ compute.

Jape goes at it a little more slowly. In particular, the definedness conditions are out in the open. Fig. 15.1 shows its first step: as usual, it's had to insert a consequence(L) step; as usual, there's a variable-distinctness proviso at the bottom; as usual, it's tried to calculate the definedness conditions out of sight, but this time there is something to be seen. The precondition on line 2 requires $0 \leq i < length(a)$, the relict of ($a[i]$ computes) $\wedge$ ($a[i] + 1$ computes).

1:	a[i]=2	assumption
2:	0≤i∧i<length(a)	bounded 1
3:	0≤i	∧ elim 2
4:	i<length(a)	∧ elim 2
5:	2+1=3	obviously
6:	a[i]+1=3	equality-substitution 1,5
7:	(a⊕i↦a[i]+1)[i]=3	index(=) 6
8:	(a⊕i↦a[i]+1)[i]=3∧0≤i∧i<length(a)	∧ intro 7,3,4
9:	a[i]=2→(a⊕i↦a[i]+1)[i]=3∧0≤i∧i<length(a)	→ intro 1-8
10:	{(a⊕i↦a[i]+1)[i]=3∧0≤i∧i<length(a)}(a[i]:=a[i]+1){a[i]=3}	array-element-assignment
11:	{a[i]=2}(a[i]:=a[i]+1){a[i]=3}	consequence(L) 9,10

Provided:
DISTINCT a, i

Fig. 15.2 Incrementing an array element: verification condition solved

$$\{\exists x(0 \le x < length(a) \land a[x] = 0)\}$$
$$i := 0$$
$$\{0 \le i < length(a) \land \exists x(i \le x < length(a) \land a[x] = 0)\}$$
while $a[i] \neq 0$ do $i := i + 1$ od
$$\{a[i] = 0\}$$

Fig. 15.3 Searching for a zero element

The steps required to establish the verification condition, shown in Fig. 15.2, correspond to the informal argument: the boundedness condition is extracted on line 2; the element-value problem is simplified on line 6 and reduced to the obvious on line 5; the rest is just straightforward application of the relevant rules. It all works!

15.4.2 Finding a zero element. Suppose there is a zero element in an array a: that is,

$$\exists x(0 \le x < length(a) \land a[x] = 0)$$

Can we write a program to find it? Of course we can! Fig. 15.3 is that program. And it's provable: Fig. 15.4 shows how the proof begins. Line 1 (invariant is established) looks fairly easy; line 2 (loop guard doesn't crash) is trivial; lines 4 and 6 (measure checks) are standard, once I filled in the measure, $\mathrm{length}(a) - i$, in place of Jape's unknown. The only difficult-looking bit is line 3 (invariant is maintained).

```
   ...
1: {∃x.(0≤x∧x<length(a)∧a[x]=0)}(i:=0)
   {0≤i∧i<length(a)∧∃x.(i≤x∧x<length(a)∧a[x]=0)}
   ...
2: 0≤i∧i<length(a)∧∃x.(i≤x∧x<length(a)∧a[x]=0)→0≤i∧i<length(a)
   ...
3: {0≤i∧i<length(a)∧∃x.(i≤x∧x<length(a)∧a[x]=0)∧a[i]≠0}
   (i:=i+1){0≤i∧i<length(a)∧∃x.(i≤x∧x<length(a)∧a[x]=0)}
   ...
4: 0≤i∧i<length(a)∧∃x.(i≤x∧x<length(a)∧a[x]=0)∧a[i]≠0→_M>0
5: integer Km                                                       assumption
   ...
6: {0≤i∧i<length(a)∧∃x.(i≤x∧x<length(a)∧a[x]=0)∧a[i]≠0∧_M=Km}
   (i:=i+1){_M<Km}
   {0≤i∧i<length(a)∧∃x.(i≤x∧x<length(a)∧a[x]=0)}
7: while a[i]≠0 do i:=i+1 od                                        while 2,3,4,5-6
   {0≤i∧i<length(a)∧∃x.(i≤x∧x<length(a)∧a[x]=0)∧¬(a[i]≠0)}
   ...
8: 0≤i∧i<length(a)∧∃x.(i≤x∧x<length(a)∧a[x]=0)∧¬(a[i]≠0)→a[i]=0
9: {0≤i∧i<length(a)∧∃x.(i≤x∧x<length(a)∧a[x]=0)}                    consequence(R) 7,8
   while a[i]≠0 do i:=i+1 od{a[i]=0}
   {∃x.(0≤x∧x<length(a)∧a[x]=0)}(i:=0)
10: {0≤i∧i<length(a)∧∃x.(i≤x∧x<length(a)∧a[x]=0)}                   Ntuple 1,9
   while a[i]≠0 do i:=i+1 od{a[i]=0}
-----------------------------------------------------------------------------------
Provided:
DISTINCT a, i
```

Fig. 15.4 Searching for zero: verification conditions

Even that bit isn't very hard, given that you have a proof calculator and a bit of experience. The formal proof is shown in Fig. 15.5. The only tricky bit is the argument by contradiction (lines 18–22) which shows that the position *i1* at which the zero occurs must be beyond the position i because $a[i] \neq 0$. The rest of the proof is straightforward, though there are some 'obviously' steps that a more arithmetically capable tool might deal with. Once more it's gratifying that all that logical machinery really works in practice.

Actually line 1 of Fig. 15.4 isn't completely obvious. It boils down to the claim that if there is an x in the range $0..length(a) - 1$ such that $a[x] = 0$, then $length(a)$ must be greater than zero. That makes sense informally, and it works out logically and arithmetically too. You might like to try to prove it.

1:	$0 \le i \wedge i < length(a) \wedge \exists x.(i \le x \wedge x < length(a) \wedge a[x]=0) \wedge a[i] \ne 0$	assumption
2:	$0 \le i$	$\wedge$ elim 1
3:	$\exists x.(i \le x \wedge x < length(a) \wedge a[x]=0)$	$\wedge$ elim 1
4:	$a[i] \ne 0$	$\wedge$ elim 1
5:	integer i1	assumption
6:	$i \le i1 \wedge i1 < length(a) \wedge a[i1]=0$	assumption
7:	$i \le i1$	$\wedge$ elim 6
8:	$i < i1 \vee i=i1$	$A \le B \triangleq A<B \vee A=B$ 7
9:	$i1 < length(a)$	$\wedge$ elim 6
10:	$a[i1]=0$	$\wedge$ elim 6
11:	$i < i1$	assumption
12:	$0 \le i+1$	obviously, from 2
13:	$i+1 < length(a)$	obviously, from 11,9
14:	$i+1 \le i1$	obviously, from 11
15:	$i+1 \le i1 \wedge i1 < length(a) \wedge a[i1]=0$	$\wedge$ intro 14,9,10
16:	$\exists x.(i+1 \le x \wedge x < length(a) \wedge a[x]=0)$	$\exists$ intro 15
17:	$0 \le i+1 \wedge i+1 < length(a) \wedge \exists x.(i+1 \le x \wedge x < length(a) \wedge a[x]=0)$	$\wedge$ intro 12,13,16
18:	$i=i1$	assumption
19:	$\neg(a[i]=0)$	$A \ne B \triangleq \neg(A=B)$ 4
20:	$a[i]=0$	equality-substitution 18,10
21:	$\perp$	$\neg$ elim 20,19
22:	$0 \le i+1 \wedge i+1 < length(a) \wedge \exists x.(i+1 \le x \wedge x < length(a) \wedge a[x]=0)$	contra (constructive) 21
23:	$0 \le i+1 \wedge i+1 < length(a) \wedge \exists x.(i+1 \le x \wedge x < length(a) \wedge a[x]=0)$	$\vee$ elim 8,11-17,18-22
24:	$0 \le i+1 \wedge i+1 < length(a) \wedge \exists x.(i+1 \le x \wedge x < length(a) \wedge a[x]=0)$	$\exists$ elim 3,5-23
25:	$0 \le i \wedge i < length(a) \wedge \exists x.(i \le x \wedge x < length(a) \wedge a[x]=0) \wedge a[i] \ne 0 \rightarrow 0 \le i+1 \wedge i+1 < length(a) \wedge \exists x.(i+1 \le x \wedge x < length(a) \wedge a[x]=0)$	$\rightarrow$ intro 1-24
26:	$\{0 \le i+1 \wedge i+1 < length(a) \wedge \exists x.(i+1 \le x \wedge x < length(a) \wedge a[x]=0)\}$ $(i:=i+1)\{0 \le i \wedge i < length(a) \wedge \exists x.(i \le x \wedge x < length(a) \wedge a[x]=0)\}$	variable-assignment
27:	$\{0 \le i \wedge i < length(a) \wedge \exists x.(i \le x \wedge x < length(a) \wedge a[x]=0) \wedge a[i] \ne 0\}$ $(i:=i+1)\{0 \le i \wedge i < length(a) \wedge \exists x.(i \le x \wedge x < length(a) \wedge a[x]=0)\}$	consequence(L) 25,26

Provided:
DISTINCT a, i

Fig. 15.5 Searching for zero: invariant is maintained

$$\{\exists x(0 \le x < length(s) \wedge s[x] = 0 \wedge x < length(buf))\}$$
$$i := 0$$
$$\left\{\begin{array}{l} 0 \le i < length(buf) \wedge \\ \exists x(i \le x < length(s) \wedge s[x] = 0 \wedge x < length(buf)) \wedge \\ \forall y(0 \le y < i \rightarrow buf[y] = s[y]) \end{array}\right\}$$
$$\textsf{while } s[i] \neq 0 \textsf{ do } buf[i] := s[i]; i := i + 1 \textsf{ od}$$
$$\{0 \le i < length(buf) \wedge s[i] = 0 \wedge \forall y(0 \le y < i \rightarrow buf[y] = s[y])\}$$
$$buf[i] := 0$$
$$\{s[i] = 0 \wedge \forall y(0 \le y \le i \rightarrow buf[y] = s[y])\}$$

Fig. 15.6 A sharp chisel doing its work

15.4.3 Buffer overflow vanquished? The buffer overflow program of Fig. 12.2 is repeated in Fig. 15.6, this time with precondition, postcondition and an invariant. The measure of the loop is $length(buf) - i$. The annotations could say more about initial and final values (s is unchanged, for example), which would make the proof longer but not any harder. It all works **provided** that there is a zero in s at a position which is within the bounds of buf (the array s can be bigger than buf and all will be well, just provided the zero is not too far along!).

This problem is now within range. It needs no more proof effort than searching for zero does. One buffer overflow problem is overthrown, at least. The crackers can begin to pack their bags.

15.4.4 Oh no it isn't! Buffer overflows are not so easy to squash. A proof of Fig. 15.6 is straightforward, the sort of thing that a compiler could knock up given a few sensible hints about the arithmetical joints. If it was that easy to block the crackers, it would have happened years ago.

The precondition of Fig. 15.6 — the assumption we rely on — is that the zero is in place and in a position which won't overflow buf. Given that precondition, the program won't bite you; without it, all bets are off. But how can we know that it's true? In most real-world situations we'd need another loop to find the zero's position and check it's within the index limits of buf. That would push the cost of the whole enterprise close to that of running a policed loop in which each index is checked before it's used, and it would add the complication of having to decide what to do when the zero's position is out of range. Real Programmers won't use policed loops, even though they understand them, and on the whole they don't understand program logic. Imagine trying to persuade them both to pay the cost of policing and also to rely on proof!

Hoare logic is not a ready-made solution to buffer overflow. We need other approaches which persuade Real Programmers to take fewer risks. Maybe.

$$
\{\ 0 \leq m < n \leq length(a) \land \exists x(m \leq x < n \land a[x] = p)\ \}
$$

```
i := m; j := n; done := ⊥;
while ¬done do
    while a[i] < p do i := i + 1 od;
    while a[j − 1] > p do j := j − 1 od;
    if i + 1 < j then
        j := j − 1;
        t := a[i]; a[i] := a[j]; a[j] := t;
        i := i + 1
    else
        done := ⊤
    fi
od
```

$$
\left\{ \begin{array}{l} 0 \leq m \leq i \leq j \leq n \leq length(a) \land i < n \land m < j \land \\ \forall xl(m \leq xl < i \rightarrow a[xl] \leq p) \land \forall xh(j \leq xh < n \rightarrow a[xh] \geq p) \land \\ (i = j \lor (i + 1 = j \land a[i] = p)) \end{array} \right\}
$$

Fig. 15.7 An efficient partition program

15.5 Programs people actually write

If program logic is to make an impact on programmers' lives, it has to be built into programming language systems and it has to work with the kind of programs that people actually write. Programmers pull all kinds of tricks to make their programs faster, smaller or sometimes just more obscure, as in the splendid annual Obfuscated C contest. Programmers are often rather clever, and their programs often work even though it seems they might be too dangerous to use.

Fig. 15.7 is an example of a program that doesn't look as if it should work but does.[2] It's a version of the partition phase of Hoare's Quicksort, one of the fastest sorting algorithms known. Given a **pivot value** p, its job is to rearrange an array segment $a[m..n-1]$ so that about half the elements, those $(\leq p)$, are in the lower part of the segment, and the other part is filled with values $(\geq p)$. Its most important feature is that it doesn't do unnecessary tests in while loop guards. In particular, there's only one bounds check per execution of the loop: $i + 1 < j$ in the choice command.

The precondition for this marvel requires only a non-empty input segment which contains p. The postcondition claims that i hasn't overtaken j $(i \leq j)$, that neither of the partitions $m..i-1$ and $j..n-1$ occupies the whole input segment $(i < n \land m < j)$, that the two segments do contain $(\leq p)$ and $(\geq p)$ values

[2] Sharp-eyed Java, C and C++ programmers will spot that this is a translation from an original which used `while (true) {...}` and a `break` command in place of $done := \top$. It would be possible to design a program logic for that kind of program but it doesn't matter: the burden of proof would be almost identical.

respectively (the xl and xh quantifications) and that either the partitions touch ($i = j$) or they are separated by a single occurrence of p ($i + 1 = j \wedge a[i] = p$). All that being so, it's safe after the partition step to sort the two partitions separately: that's done with a couple of recursive calls which we needn't discuss. (There is also an important omission: I haven't attempted to specify that a always contains a permutation of its input value. That's essential when specifying a sorting algorithm, and it's obviously true in this example because all it does is exchange elements of a, but I wanted to concentrate on the bounds checks.)

I expect the program looks as dangerous to you as it once did to me. The internal while loops look as if they might run off the end of the array; the exchange looks as if it might not always put things in the right places. It's all written for speed. (It would be faster still if it didn't use $j - 1$ as an index quite so often, but for simplicity I've ignored that wrinkle.)

Fig. 15.8 shows invariants and some intermediate assertions. I've inserted a skip to make the body of the loop an Ntuple, so that I can include an initial assertion before the first internal while. The invariant requires

$$\exists yl(m \leq yl < j \wedge a[yl] \leq p) \wedge \exists yh(i \leq yh < n \wedge a[yh] \geq p)$$

— there is an element in the segment below j which will stop the $j := j-1$ loop, and one at or above i which will stop the $i := i+1$ loop. Proofs that the internal loops don't run off the end of the segment are then just like the zero-searching example (Fig. 15.3), and at the same time it would seem easy to prove that they maintain the partitioning quantifications

$$\forall xl(m \leq xl < i \rightarrow a[xl] \leq p) \wedge \forall xh(j \leq xh < n \rightarrow a[xh] \geq p)$$

The hardest bit of the proof is to show that the interchange program which swaps $a[i]$ and $a[j - 1]$ preserves the invariant, setting things up for the next time round. The verification condition part of that proof is shown in Fig. 15.5, produced by applying the choice rule to the invariant and pushing backwards through all the assignments. It contains several occurrences of the updatable-sequence formula

$$(a \oplus i \mapsto a[j-1] \oplus j - 1 \mapsto a[i]) \tag{15.1}$$

so it seems to have done the right exchange.

Lines 3, 4 and 5 are pretty straightforward. Line 6 requires us to pick an element of the updated sequence which is $(\leq p)$; we know (line 1.3) that $a[j-1] \leq p$, so obviously the index to pick for the updated sequence is i. An $\exists$ intro step, a couple of applications of the indexing rules, plus the fact ($i + 1 < j$) that $i \neq j - 1$ and it's done. Line 7 is similar, this time picking $j - 1$ as the index. Line 8 requires the loop-rolling theorem (Fig. 14.12 page 215), then $\forall$ intro, and finally an unpicking of the indexing formula given that $xl < i$. Line 9 is similar.

$$\{\ 0 \leq m < n \leq length(a) \wedge \exists x(m \leq x < n \wedge a[x] = p)\ \}$$
$$(i := m; j := n; done := \bot)$$
$$\left\{\begin{array}{l} 0 \leq m \leq i \leq j \leq n \leq length(a) \wedge \\ \exists yl(m \leq yl < j \wedge a[yl] \leq p) \wedge \exists yh(i \leq yh < n \wedge a[yh] \geq p) \wedge \\ \forall xl(m \leq xl < i \rightarrow a[xl] \leq p) \wedge \forall xh(j \leq xh < n \rightarrow a[xh] \geq p) \wedge \\ (done \rightarrow i = j \vee (i+1 = j \wedge a[i] = p)) \end{array}\right\}$$

while $\neg done$ do

skip

$$\left\{\begin{array}{l} 0 \leq m \leq i \leq j \leq n \wedge n \leq length(a) \wedge \\ \exists yl(m \leq yl < j \wedge a[yl] \leq p) \wedge \exists yh(i \leq yh < n \wedge a[yh] \geq p) \wedge \\ \forall xl(m \leq xl < i \rightarrow a[xl] \leq p) \wedge \forall xh(j \leq xh < n \rightarrow a[xh] \geq p) \wedge \\ \neg done \end{array}\right\}$$

while $a[i] < p$ do $i := i + 1$ od

$$\left\{\begin{array}{l} 0 \leq m \leq i \leq j \leq n \wedge n \leq length(a) \wedge \\ \exists yl(m \leq yl < j \wedge a[yl] \leq p) \wedge \\ \forall xl(m \leq xl < i \rightarrow a[xl] \leq p) \wedge \forall xh(j \leq xh < n \rightarrow a[xh] \geq p) \wedge \\ \neg done \wedge a[i] \geq p \end{array}\right\}$$

while $a[j-1] > p$ do $j := j - 1$ od

$$\left\{\begin{array}{l} 0 \leq m \leq i \leq j \leq n \wedge n \leq length(a) \wedge \\ \forall xl(m \leq xl < i \rightarrow a[xl] \leq p) \wedge \forall xh(j \leq xh < n \rightarrow a[xh] \geq p) \wedge \\ \neg done \wedge a[i] \geq p \wedge a[j-1] \leq p \end{array}\right\}$$

if $i + 1 < j$ then

$j := j - 1;$

$t := a[i]; a[i] := a[j]; a[j] := t;$

$i := i + 1$

else

$done := \top$

fi

od

$$\left\{\begin{array}{l} 0 \leq m \leq i \leq j \leq n \leq length(a) \wedge i < n \wedge m < j \wedge \\ \forall xl(m \leq xl < i \rightarrow a[xl] \leq p) \wedge \forall xh(j \leq xh < n \rightarrow a[xh] \geq p) \wedge \\ (i = j \vee (i+1 = j \wedge a[i] = p)) \end{array}\right\}$$

Fig. 15.8 Annotated partition program

Lines 11 and 12 come from premise 1.2 and $\wedge$ elim; lines 13 and 14 from premise 1.3.

So it can be done, and the most important observation to make is that it would take no particular ingenuity to make the proof. A mechanical search could do it. Maybe the Verifying Compiler, that vision of a program processor which checks logical claims the way that present-day compilers check typing claims, is not so far off after all.

1: i+1<j, a[j-1]≤p, a[i]≥p, ¬done, ∀xh.(j≤xh∧xh<n→a[xh]≥p) premises

2: ∀xl.(m≤xl∧xl<i→a[xl]≤p), n≤length(a), j≤n, i≤j, m≤i, 0≤m premises

...

3: m≤i+1

...

4: i+1≤j-1

...

5: j-1≤n

...

6: ∃yl.(m≤yl∧yl<j-1∧(a⊕i↦a[j-1]⊕j-1↦a[i])[yl]≤p)

...

7: ∃yh.(i+1≤yh∧yh<n∧(a⊕i↦a[j-1]⊕j-1↦a[i])[yh]≥p)

...

8: ∀xl.(m≤xl∧xl<i+1→(a⊕i↦a[j-1]⊕j-1↦a[i])[xl]≤p)

...

9: ∀xh.(j-1≤xh∧xh<n→(a⊕i↦a[j-1]⊕j-1↦a[i])[xh]≥p)

...

10: done→i+1=j-1∨(i+1+1=j-1∧(a⊕i↦a[j-1]⊕j-1↦a[i])[i+1]=p)

...

11: 0≤j-1

...

12: j-1<length(a)

...

13: 0≤i

...

14: i<length(a)

15: 0≤m∧m≤i+1∧i+1≤j-1∧j-1≤n∧n≤length(a)∧∃yl.(m≤yl∧yl<j-1∧(a⊕i↦a[j-1]⊕j-1↦a[i])[yl]≤p)
∧∃yh.(i+1≤yh∧yh<n∧(a⊕i↦a[j-1]⊕j-1↦a[i])[yh]≥p)∧∀xl.(m≤xl∧xl<i+1→(a⊕i↦a[j-1]⊕j-1↦a[i])[xl]≤p)
∧∀xh.(j-1≤xh∧xh<n→(a⊕i↦a[j-1]⊕j-1↦a[i])[xh]≥p)
∧(done→i+1=j-1∨(i+1+1=j-1∧(a⊕i↦a[j-1]⊕j-1↦a[i])[i+1]=p))∧0≤j-1∧j-1<length(a)∧0≤i∧i<length(a)   ∧ intro 2.6,3,4,5,2.2,6,7,8,9,10,11,12,13,14

Fig. 15.9 Treating the verification condition for the interchange section

15.6 Do verified programs run faster?

Novice programmers are rightly terrified of programs like Fig. 15.7: they already know that array-indexing loops can overrun their bounds, and those internal whiles don't test for that possibility. In my youth I wouldn't have dared to run the program, let alone write it. But now I know that under very easily established circumstances — a non-empty input segment and an assignment like $p := a[(m+n) \div 2]$ — it works perfectly and those loops don't need to test the array bounds. The program is safe without the checks and it's faster too. Safe programs which are also faster: we can have the cake **and** keep the penny! Maybe that will persuade Real Tough Programmers to listen.

Maybe not. Don't forget the buffer overflow example: we don't have a panacea. Arithmetic reasoning is still a problem: we can automate lots of it but we'd have to ask programmers to give us hints to help us past the hard bits, and those hints would be another source of mistakes. So we're not there yet, not by a long way.

The dream, though, is to automate this stuff, build it into compilers and such, and find a way to make the arithmetic hinting job something that ordinary mortals can deal with. That work is going on (and, in fact, we have justified ambitions way beyond anything in this book). It will probably last your lifetime if you want to join in. Careless use of dangerous programs is hurting us right now. We have to help the Real Tough Programmers with their Real Tough Problems. I believe we're beginning to succeed.

15.7 There are lots more examples

Whether or not we win the RTPs' RTP battle for them, program proof is still fun. There are more array-program examples in Jape: try them.

Index